The Handbook of Lithium-Ion Battery Pack Design

The Handbook of Lithium-Ion Battery Pack Design

Chemistry, Components, Types and Terminology

John Warner

Grand Blanc, MI, USA

ELSEVIER

AMSTERDAM • BOSTON • HEIDELBERG • LONDON • NEW YORK • OXFORD
PARIS • SAN DIEGO • SAN FRANCISCO • SINGAPORE • SYDNEY • TOKYO

Elsevier
Radarweg 29, PO Box 211, 1000 AE Amsterdam, Netherlands
The Boulevard, Langford Lane, Kidlington, Oxford OX5 1GB, UK
225 Wyman Street, Waltham, MA 02451, USA

ISBN: 978-0-12-801456-1

British Library Cataloguing in Publication Data
A catalogue record for this book is available from the British Library

Library of Congress Cataloging-in-Publication Data
A catalog record for this book is available from the Library of Congress

For Information on all Elsevier publications
visit our website at http://store.elsevier.com/

Contents

Figure Captions

Chapter 2

Chapter 3

Chapter 4

Chapter 5

Chapter 6

Chapter 7

Chapter 8

Chapter 9

Chapter 10

Chapter 12

Chapter 14

Chapter 15

Preface

In early 2009 as the US automotive industry was in the midst of its restructuring, I took advantage of the changing industry to join a new energy start-up and enter into the lithium-ion battery space. As I worked to make the transition from a major OEM to the lithium-ion battery industry, I purchased pretty much every book I could find on lithium-ion batteries looking for one that gave me the basic information, which I would need to be successful. However, I found that while there were some good books on the market, they were either very technical and targeted at engineers or focused on different markets such as laptops or on different technologies such as Nickel-Metal Hydride batteries. Over the following years I spent a lot of time working with chemists, engineers, and battery scientists to learn as much as I could.

However, about 18 months ago I was speaking to a colleague who was asked a simple question as we were coming out of a meeting—how do you know all of this stuff? And it occurred to me that there may be a need for a tool to help people, who are working in the lithium-ion battery industry but are not battery experts but who still need to gain a better understanding of the industry and the products in order to do their jobs better. In other words, I decided to write the book that I wish I had when I started in the industry. I spent a lot of time thinking back to that time and the types of questions that I had. As someone who had spent a career in engineering organizations but was not an engineer, I had to ask a lot of questions and made a lot of crib notes but I could not find a single source for everything I was trying to learn.

This book is the realization of that knowledge gap in the industry. This book is intended for everyone; you do not have to be an engineer in order to gain an understanding of batteries. In fact, as the battery industry has grown so much over the past 10 years, there have been a lot of new people coming into the battery world from other industries. That means a lot of people with great experience in their various specialties who now need to learn about lithium-ion batteries. Maybe you are a student in one of the new energy storage system programs that are beginning to sprout up in universities across the world, or perhaps you are a purchasing manager who is now tasked with buying a whole new set of components and have no idea what it is they are, or what if you are a thermal engineer who has now moved into the battery world—this book is for you!

In this book you will first begin by gaining an understanding of the history of batteries, as we do not want to repeat any of the mistakes of the past; it is important to understand what has come before us. The next most challenging part of moving into a new industry is understanding the lingo; this book will also help to give you that basic understanding. You will also be able to gain an understanding of the basic math that can be used in doing the first run sizing of a battery—this section is the result of 7 years of taking notes and back calculating some of the work I had seen engineers do over the years. This is followed by chapters that will introduce you to the different parts of the battery, the industry organizations that are out there, and a wide range of different applications that are being powered by batteries—some lithium-ion and some with other technologies.

So whether you are looking to learn something about one aspect of lithium-ion batteries to bolster your knowledge or are entirely new and are looking to learn all about the basics, this book will be a good tool to add to your toolbox.

I have come to believe that, after nearly 2 years of writing and editing this book, it is an ongoing project that will continue to evolve as the industry and technologies themselves evolve. Lithium-ion battery technology is not fixed; it is a constantly evolving field with new innovations, inventions, and chemistries emerging almost daily. This book will give you a great basis to continue your education. So charge up and get started!

Acknowledgments

I would like to begin by thanking my wife Amy and my children Erika and Lukas for their support and encouragement while I have undertaken this project. Without your patience and support over the many weekends and evenings, this book would not have been possible.

I also wish to thank the following people for their contributions to my inspiration and knowledge and other help in creating this book: Bob Purcell, whose background in the automotive electrification field proved very helpful in offering direction and insights as the project was being outlined; Bob Galyen, who has spent his career as a leader in the battery-energy storage industry who provided constant encouragement and support as well as providing direction when needed; Dr Per Onnerud and Dr Christina Lampe-Onnerud, who both became early mentors to me in the lithium-ion battery world; Dr JR Lina, who took me under his wing early in my battery career and taught me many of the basic and key concepts around which this book is based; Subhash Dhar, who has led more energy storage companies over the past 20+ years than anyone else I know; Jon Bereisa, who has been involved in electrification throughout many aspects of his career and is perhaps one of the best resources I have known who can speak on just about any topic. And I need to include a special thanks to some of the people who were initial reviewers of the idea for this book, which helped me to guide the direction and scope of it including Bob Kruse, Dell Crouch, Lori Hutton, Oliver Gross, and many others.

I apologize if I missed anyone, but everyone I have worked with and had interactions with over the past 7 years or more have all been the inspiration for this book and I thank you!

Acronyms List

A	Ampere
AC	Alternating Current
AGM	Absorbed Glass Mat
Ah	Ampere hour
AIAG	The Automotive Industry Action Group
ALBAC	Advanced Lead Acid Battery Council
ARB	Air Resource Board
ASIC	Application Specific Integrated Circuit
ASQ	American Society for Quality
AUV	Autonomous Underwater Vehicle
BCI	Battery Council International
BDU	Battery Disconnect Unit
BEV	Battery Electric Vehicle
BMS	Battery Management System
BOL	Beginning of Life
CAD	Computer-aided Design
CAE	Computer-aided Engineering
CAEBAT	Computer-aided Engineering for Electric-Drive Vehicle Batteries
CAFE	Corporate Average Fuel Economy
CARB	California Air Resource Board
CATARC	China Automotive Technology and Research Center
CES	Community Energy Storage
CFD	Computational Fluid Dynamics
CID	Current Interrupt Device
CSC	Cell Supervision Circuit
DC	Direct Current
DEC	Diethyl Carbonate
DES	Distributed Energy Storage
DFMEA	Design Failure Modes Effect Analysis
DFR	Design for Reliability
DFS	Design for Service

DFSS	Design for Six Sigma
DMC	Dimethyl Carbonate
DOD	Depth of Discharge
DOE	U.S. Department of Energy
DOE	Design of Experiments
DVP&R	Design, Validation Plan & Report
EC	Ethylene Carbonate
ECSS	Electrochemical Storage System
eMPG	Electric Miles per Gallon
EDV	Electric Drive Vehicles
EES	Electrochemical Energy Storage
EMC	Ethylmethyl Carbonate
EMC	Electromagnetic Compatibility
EMI	Electromagnetic Interference
EMS	Energy Management System
EOL	End of Life
EREV	Extended Range Electric Vehicle
ESS	Energy Storage System
EUCAR	European Council for Automotive Research and Development
EV	Electric Vehicle
EVAA	Electric Vehicle Association of America
FCEV	Fuel Cell Electric Vehicle
FEA	Finite Element Analysis
FMEA	Failure Modes Effect Analysis
GEO	Geosynchronous Earth Orbit
GEV	Grid-tied Electric Vehicle
HC	Hydrocarbon
HEO	High Earth Orbit
HEV	Hybrid Electric Vehicle
HD	Heavy Duty
HIL	Hardware in the Loop
HPDC	High Pressure Die Cast
HPPC	Hybrid Power Pulse Characterization
HV	High Voltage
HVAC	Heating Ventilation Air Conditioning
HVFE	High Voltage Front End
HVIL	High Voltage Interlock Loop
IBESA	International Battery and Energy Storage Alliance
ICB	Interconnect Board
ICE	Internal Combustion Engine

IEC	International Electrotechnical Commission
IEEE	Institute of Electrical and Electronics Engineers
INL	Idaho National Laboratory
IP	International Protection
IP	Ingress Protection
IPVEA	International Photovoltaic Equipment Association
ISO	International Organization on Standardization
kWh	kilo-watt hour
LAB	Lead Acid Battery
LCO	Lithium-ion Cobalt Oxide
LD	Light Duty
LEO	Low Earth Orbit
LEV	Low Emissions Vehicle
LEV	Light Electric Vehicle
LEVA	Light Electric Vehicle Association
LFP	Lithium-ion Iron Phosphate
LIB	Lithium-ion Battery
LIP	Lithium-Ion Polymer
LiPo	Lithium-Ion Polymer
LI-Poly	Lithium-Ion Polymer
LMO	Lithium-ion Manganese Oxide
LPG	Liquid Propane Gas
LTO	Lithium-ion Titanate Oxide
LV	Low Voltage
MEO	Medium Earth Orbit
μHEV	Micro Hybrid Electric Vehicle
MPG	Miles per Gallon
MSD	Manual Service Disconnect
MTBF	Mean Time between Failures
MTTF	Mean Time to Failure
MY	Model Year
MWh	Mega-watt hour
NAATBatt	National Association for Advanced Technology Batteries
NCA	Lithium-ion Cobalt Aluminum
NEMA	National Electrical Manufacturers Association
NEV	Neighborhood Electric Vehicle
NEV	New Energy Vehicle (China)
NHTSA	National Highway Transportation Safety Administration
NiCd	Nickel Cadmium
NiMh	Nickel Metal Hydride

NMC	Lithium-ion Nickel Manganese Cobalt
NREL	National Renewables Energy Laboratory
NTC	Negative Thermal Coefficient
NTCAS	National Technical Committee on Automotive Standardization (China)
OEM	Original Equipment Manufacturer
ORNL	Oak Ridge National Laboratory
OSV	Off-Shore Vessel
PbA	Lead Acid
PCB	Printed Circuit Board
PCM	Phase Change Material
PE	Polyethylene
PFMEA	Process Failure Modes Effect Analysis
PHEV	Plug-In Hybrid Electric Vehicle
PMS	Power Management System
PNNL	Pacific Northwest National Laboratory
PP	Polypropylene
PRBA	Portable Rechargeable Battery Association
PSV	Platform Supply Vessel
PTC	Positive Thermal Coefficient
PV	Photovoltaic
PVDF	Polyvinylidene Fluoride
PZEV	Partial Zero Emissions Vehicle
REEV	Range Extended Electric Vehicle
RESS	Rechargeable Energy Storage System
REX	Range Extender
SAC	Standardization Administration of China
SAE	Society of Automotive Engineers
SEI	Solid Electrolyte Interphase
SIL	Software in the Loop
SLA	Standard Lead Acid
SLI	Starting, Lighting, Ignition
SNL	Sandia National Lab
SOC	State of Charge
SOH	State of Health
SOL	State of Life
SRU	Smallest Replaceable Unit
S/S	Stop/Start
T&D	Transmission & Distribution
TMS	Thermal Management System
TTF	Test to Failure

UAV	Unmanned Aerial Vehicles
UL	Underwriter's Laboratory
UN	United Nations
UPS	Uninterruptible Power Supply
USABC	U.S. Advanced Battery Consortium
USCAR	United States Center for Automotive Research
UUV	Unmanned Underwater Vehicles
VDA	Verband der Automobilindustrie
VRLA	Valve Regulated Lead Acid
VOC	Voice of the Customer
VTB	Voltage, Temperature monitoring Board
VTM	Voltage, Temperature Monitoring
W	Watt
W/kg	Watt per kilogram
W/L	Watt per liter
Wh	Watt-hour
Wh/kg	Watt-hour per kilogram
Wh/L	Watt-hour per liter
ZEV	Zero Emissions Vehicle

CHAPTER 1

Introduction

Today lithium-ion (Li-ion) batteries are everywhere…they power our watches, smart phones, tablets, laptops, portable appliances, GPS devices, handheld games, and just about everything else we carry with us today. But they are also beginning to power our neighborhoods, our homes, and our vehicles, or perhaps when talking about transportation applications, it is more accurate to say that batteries power our transportation *again*. And today as these industries continue to experience rapid growth, many people who have not previously worked with Li-ion batteries now find themselves in the role of a business professional, technician, or engineer who is moving into the field of Li-ion batteries and are in need of an introduction to Li-ion battery technology. In that case, this book is designed specifically for you!

This book is intended to introduce a variety of topics that surround Li-ion batteries and battery design at a detailed enough level to make batteries understandable for the "layman." If you are an engineer, you will swiftly understand these concepts. However, if you are like many of us and are not an engineer, then this book will help you make sense of the world of Li-ion batteries and be able to speak intelligently about them. The concepts in this book are focused on vehicle electrification, but are also relevant to many other applications including stationary energy storage, marine and offshore vessels, industrial motive, robotics, and other types of electric applications. In essence, this book is intended to take the mystery out of modern battery applications. But let me make a disclaimer as we get started, this book is not intended to make you a battery engineer nor is it intended to replace your battery engineering team. It is instead a tool to add to your tool kit.

Batteries are unique in the field of energy storage products as they both create the energy through chemical processes and then store the energy within the same device. Other energy storage devices require the energy to be generated in one place and stored in another. For example, in an automobile, the energy is created through a refining of liquid crude oil, it is then transferred to the service stations, where it is again stored until you purchase it and store it again as a liquid fuel in the tank, it is finally converted into energy (and work) in the combustion process of the internal combustion engine.

In a Chapter I wrote for the *Handbook of Lithium-ion Battery Applications* (Warner, 2014), I offered a brief look at Li-ion battery design considerations and discussed cells, mechanical, thermal, and electronic components of Li-ion battery packs. This book will build on that initial discussion and dig deeper into each of those systems and delve into some of the formulas and calculations that are used when making battery pack decisions. Additionally, we

The Handbook of Lithium-Ion Battery Pack Design. http://dx.doi.org/10.1016/B978-0-12-801456-1.00001-4

will take a look at some of the testing and certification requirements, the growing group of industry organizations and discuss some of the various applications that are benefitting from the addition of electrical energy storage technology.

With the recent exposure that has occurred in the media surrounding Li-ion batteries, such as the Boeing 787 Dreamliner battery failures that caused the delay in launching the new plane and an intensive investigation and several battery failures in both Tesla and Chevrolet's electric vehicles (Klayman, 2013; Lowy, 2011; Santos, 2013), the focus on Li-ion battery design as a *system* has become much more important. The perspective that I have taken for some time now is that it is not the cell that makes the system safe, in fact it is exactly the opposite it is the system that makes the battery cells safe. What I mean by this is that you can take the highest quality and best performing Li-ion battery in the world and put it into a poorly designed pack and it will fail, it could suffer from reduced life, low power, and ultimately safety issues. Conversely, you can take a relatively poorly designed or manufactured cell and make it operate relatively safely with a good energy storage *System* design. Li-ion energy storage system design requires taking a systems level approach—there is no magic chemistry available that will make a pack safe.

Factors Influencing Consumer Adoption of Electric Vehicles

Another question that we should ask as we begin our discussion on Li-ion batteries is, what are the factors that will drive consumer adoption of Li-ion battery powered vehicles? There are effectively five factors that need continuous attention by all of the major stakeholders in order to ensure the successful growth of battery powered vehicles and energy storage systems. These factors include:

1. Cost—great progress has already been made in reducing battery costs, but there is still much work to be done to make Li-ion-based battery solutions affordable enough to enable high-volume applications. And when it comes to stationary energy storage systems, there are many competing technologies, making it even more important for the costs of Li-ion to continue to drop.
2. Availability—there are more new electrified vehicle applications available to the consumer every year, but when we look at the more highly electrified plug-in hybrid electric vehicle and electric vehicles, there are still a relatively limited amount of products available to the consumer.
3. Range anxiety—even with more electrified versions of current vehicles available, there is still an area of customer concern around the electric range that the vehicles can achieve.
4. Education—perhaps one of the areas that neither the industry nor the governments have done a very good job around is in educating the consumer on the differences, similarities, benefits, and challenges of electrified vehicles. Consumers still consider any level of electrification to be "high technology" rather than mass market ready. But in all reality, the "conventional" vehicle is already almost entirely electrified outside of the powertrain.

5. Charging Infrastructure—this factor is somewhat of a chicken and the egg scenario, no one wants to invest in a charging network that does not get used, but without a charging network consumers are not likely to purchase vehicles that need to use them.

These are, in my opinion, the five keys issues that need to be resolved in order to ensure that electrification technology will get more deeply integrated into the transportation and stationary base architectures. Some of these, such as cost, are hurdles that will be achieved through increased volumes in the applications as well as in advancements in technology. But cost and volume are directly correlated to consumer awareness and having a charging infrastructure that matches consumer demand, along with having a higher number of available electrified applications.

If we think of the typical automotive consumer, we find that we have become much more dependent on the technology of our vehicles but at the same time our understanding of how that technology works has declined. In other words, most consumers do not care or even understand what is under the hood, only that it will get them everywhere they need to go. For the early adopter consumer, factors such as the "green" factor and the environmental factor tend to be important purchase factors.

But for the mass market consumer, these become what I refer to as hygiene factors. The term is taken from the field of organizational psychology and motivation, where it was first used by Frederick Herzberg. Herzberg described the phrase hygiene factor in terms of motivation as a factor that if absent will cause dissatisfaction, but if present by themselves will not cause satisfaction (Herzberg, Mausner, & Snyderman, 1959). Batteries and battery-powered vehicles fall into this category for the mainstream consumer.

Having an advanced battery will not be a reason for a mass market consumer to purchase, but the performance improvement as seen in fuel economy will make it more competitive with other options. In order to make the shift from the early adopter to the mass market user, the technology needs to be able to offer a financial payback, internal rate of return, or significant improvements in fuel economy. In other words, the consumer needs to know that the investment will offer them a financial benefit in reduced fuel costs. In addition to this, the mass market consumer needs to know that the technology is "bullet proof," they do not want to invest in early stage technologies since there is not a lot of proof that it works. The mass market consumer is also not willing to give up any performance or range of the vehicle. This one is of extreme importance in the growth of electric vehicles, as mass market consumers will require a vehicle that can offer them the same range as the traditional vehicle.

Evolving Vehicle Technology Needs

Increasing levels of vehicle electrification such as electric heating, ventilation, air conditioning; electric power steering; electric oil pumps; electric fuel pumps; and more in-vehicle electrification, such as GPS, advanced radio, and navigation systems, all require greater levels of power. Greater levels of vehicle to vehicle communications and autonomous vehicles will also require high levels of electrification. More battery technology also enables greater optimization of the engine by transferring these parasitic loads from the engine to the battery, which in turn allows for more engine downsizing.

Yet today most electrically based vehicles are still based on architectures that were designed for internal combustion engines. In fact, in most of these vehicles, there is no standard location for a battery, which is why we see so many different installation locations within different vehicles. Today, with only minor exceptions, most electrically powered vehicles package their batteries under the seats, in the trunk, or in the transmission tunnel—in essence, trying to fit the square peg of a battery into the round hole of the existing vehicle architecture. Future vehicles will be designed with the battery as one of the design criteria so it will be an integral part of the vehicle architecture rather than being added as an afterthought.

Purpose of the Book

With all of these challenges in automotive and other energy storage applications, it is becoming more challenging to find a common means to discuss them. This book is intended to help bridge that gap by answering questions such as: What is the difference between a power battery and an energy battery? What is the difference between different Li-ion cell chemistries? Should the pack be liquid or air-cooled? What is a battery management system (BMS) and what does it do? These are the types of questions that this book will try to answer. Will this qualify you to be a battery engineer when you are done? Definitely not, but will it prepare you to ask the right questions and have a solid understanding of the topic? Absolutely.

Throughout the book I will attempt to use some analogies to correlate the terms and concepts that are used here to make them more easily understandable. For instance, we may think of the term energy, usually measured in kilowatt hours (kWh), in terms of being analogous to the size of a gas tank. So when we are speaking about the energy of a battery, we are talking about how much energy that battery "tank" holds.

I will use many examples from the automotive field to demonstrate the principles and concepts discussed here but remember that all of these concepts apply to a much wider variety of uses and applications. In Chapter 15, Li-ion Battery Applications, we will take a very wide look at the different examples of Li-ion batteries in a wide variety of applications ranging from e-bikes to electrified vehicles to forklifts, buses and stationary energy storage.

Li-ion batteries more than many other subsystems in the vehicle, require a "systems"-level approach to engineering and design. Battery pack engineering begins with the chemistry that happens at the cell level, then includes the electrical performance of both the cell and the pack, the electronics for the control system, the thermal management of the cells, the electronics and software that manage the system, and the mechanical and structural components of the battery. In other words, the single topic of Li-ion battery pack engineering involves almost every engineering field.

One of the biggest challenges that many people have when thinking about batteries is that they have a tendency to treat batteries like they are commodities, parts, and components that are easily substitutable with similar parts from different manufacturers. Unfortunately, Li-ion batteries are more akin to critical core engineering systems such as engines or transmissions. This commodity-based perspective leads to a narrow perspective for the planning, scope, and cost of the battery systems. This book will attempt to offer a view into the complexity of the battery system in order to help the reader understand why the battery is not a commodity in the automotive market today and is not likely to become one in the near future.

This leads us to the second goal of this book which is to put these different subcomponents and their discussions into the perspective of the standard engineering design process. One of the frequent challenges of engineering battery packs, especially for large and transportation-related applications, is that the timing related to these activities is not clearly understood. Many people will make comparisons to the portable power industry which uses about a 12- to 18-month-design cycle from concept to mass production for a product with a 2- to 3-year life. However, when we apply the repeating "Design, Build, Test" design process to automotive applications, it is generally a 2- to 3-year-process from concept to production. This surprises many people, however, we must remember that the typical automotive design cycle can run up to 4 years or longer depending largely on the amount of validation and verification testing that is needed.

There are several reasons for the long duration of timing, including long testing periods (a battery "life cycle" test typically runs for a year or longer), long production tooling times (some "hard" tooling can have lead times of 25–30 weeks or greater), and on top of these long lead time items. There are some "hard stops" in the process such as needing to complete testing before sourcing and achieving design freezes before beginning prototype sourcing. Add to these, the need to complete, on average, two to three design cycles (usually timed to match a vehicle prototype build) and you end up with a 24- to 36-month-long battery pack design cycle.

Of course that does not mean that there are not methods that can be used to reduce that timing largely depending on whether it is a production program, the final intended use, the lifetime performance duration, the existence of similar products, and the amount of risk that the company is willing to take. In a typical design project, all phases are completed in series, in other words one after another. However, there are various reasons that you may be able to overlap some of the phases and run them in parallel in order to reduce the overall project timing.

Chapter Outline

Chapter 2 will offer a brief history of both vehicle electrification, the evolution of the modern battery and the growth of electric infrastructure, discussing where they came from and where they are today...since this may offer us insights into where the technology could lead in the future. This chapter will provide the reader with the basic background understanding of how the industry came to be and the challenges that it has faced throughout its hundred-plus-year growth in an effort to gain an understanding of how the industry has come to where it is today.

Chapter 3 will begin introducing the terminology and basic concepts around Li-ion batteries and vehicle electrification. One of the greatest challenges that people face when they enter a new industry is having a lack of comprehension of the lingo, terminology, and phrases that are used to describe different parts of the product. In addition to gaining an understanding of the terminology used in this industry, the basic forms of electrification will be described in simple terms and their key components will be identified.

In Chapter 4, we will begin describing the process of selecting and sizing a battery for an application. This will include offering some simple formulas that can be used to help determine capacity, power, and voltage for a battery system. This chapter attempts to clearly describe the "how" of developing a battery pack concept using basic calculations based on Ohms' Law.

Chapters 5 and 6 are intended to offer supporting discussion to the engineering design process. Chapter 5 reviews the topics of Design for Reliability and Design for Service. As Li-ion batteries are relatively new in the market place, they need an extra amount of attention when it comes to reliability and service. In most cases, the Li-ion cells have not actually been in use in the market place for as long as they are being warranted. With many batteries being offered to the consumer with much longer warranty periods than the rest of the vehicle, what needs to be considered in the engineering design process to ensure the battery meets these targets? Therefore it is important to understand the factors that affect long-term quality and reliability in advanced battery solutions. Chapter 6 builds on this with a discussion on computer-aided design and analysis programs and software that are available for use. There are few software products that are available in the market today that are designed specifically for the battery designer. However, there are some software packages that can be used to evaluate thermal performance or mechanical performance of a battery.

Chapter 7 will review the various battery cell technologies that are available on the market today. This will include chemistries from lead acid to nickel metal hydride and sodium-based chemistries as well as all of the basic forms of Li-ion cells and discuss the differences between them. While the focus will be on Li-ion cells, this chapter will provide a brief review of these other chemistries and their applicability to vehicle electrification. This chapter will also discuss the criteria around Li-ion cell selection and the varieties of cells that are available.

Chapter 8 will begin looking at the control electronics side of the battery. What is a BMS and what does it do? Can a battery system operate without a BMS? What are the major components of the BMS and how do they interact with the battery and the vehicle systems? What are the different types of BMS systems and which is most appropriate for my application? These are the types of questions that will be discussed in Chapter 8.

Chapter 9 will break down the other electronic components that are needed to make a battery pack run and to make it safe. What is a contactor and a precharge contactor? What is a service disconnect and why would you need one? Both high-voltage and low-voltage electrical systems will be discussed.

Chapter 10 will review thermal management of the battery pack. How do you select the proper system? What are the criteria needed to make this decision? What types of thermal management systems are most effective for various types of batteries in different regions and applications? When is a liquid management system better than a forced air-based system?

Chapter 11 will review the topics surrounding packaging, material selection, mechanical and structural components of the battery system. In this chapter, we will examine the need for structural integrity around the battery and how to achieve it.

Chapter 12 will discuss battery abuse tolerance including what are the general operational and survival temperature ranges that Li-ion batteries can operate in. This chapter will also include a discussion on abuse, characterization, and certification testing. In particular, we will review some of the certification testing that is targeted at Li-ion batteries from the United Nations (UN) and Underwriter's Laboratory (UL).

In Chapter 13, we will review the growing industrial standards that are being enacted around the globe, looking at everything from SAE to IEEE to IEC and the UN shipping requirements and how they impact the design of modern batteries. This chapter will include voluntary standards, mandatory standards, as well as R&D and trade organizations.

Chapter 14 will review the concept around second life, remanufacturing, repair, and recycling of Li-ion batteries. This part of the industry is in its very early stages of formation and so is ripe for the right company to develop a strong market position. Today lead acid batteries have amazingly high levels of recycling, above 98% in most of the United States and Europe. What kind of infrastructure is needed in order for Li-ion batteries to achieve the same level of recycling? And what is the value of second-life batteries? Will the initial costs come down low enough to make second life a nonstarter? Or will Li-ion battery costs decrease at a slow enough rate to make second-life batteries a viable solution?

Chapter 15 will take a deeper look at the types of applications where battery power can offer significant benefits spanning from bicycles to automobiles to stationary, marine and industrial applications. The amount of potential applications that can be improved or transitioned to Li-ion is almost incalculable. Any application that has high fuel consumption or high emissions may benefit from adding an electrification component.

Finally, Chapter 16 will summarize some of the key points and more importantly discuss the future of Li-ion batteries in the transportation and energy storage markets. What are the promising technologies and chemistries that could change the market in the future? And what are the trends that are moving us closer to an electric future?

CHAPTER 2

History of Vehicle Electrification

Electricity may just be one of the most overlooked modern "conveniences" of our time. It tends to be taken for granted until we experience a power outage, then we realize just how deep-rooted it has become in our daily lives. One of the most interesting facts about electricity and, in particular, electrical energy storage and electric vehicles is that they are not entirely new ideas. In this chapter, we will review the history related to the growth of the electricity market, battery energy storage, and the electric vehicle. Here we will follow the timeline from the earliest known energy storage batteries to the late-industrial revolution when electricity became available to the masses all the way to the charge toward vehicle electrification that began in the late-1800s and continues into today.

But you may ask why should we care about the history of the electric storage battery and the electric vehicle? As has been stated many times before "Those that cannot remember the past are condemned to repeat it" (Santayana, 1905). As with any technological innovation, the energy storage battery, particularly as it relates to vehicle electrification, has had many fits and starts over the past 150 years or so. So gaining a basic understanding of how the technology has come to evolve may play an important role in understanding where it may continue to go in the future and what the key enablers are to making it truly successful today.

The adoption of a new technology does not always happen quickly and in fact may be heavily reliant on other factors. The chart shown in Figure 1 below, originally printed in the New York Times (Felton, 2008), shows new technology introductions compared to the percent of US households that have adopted them. From this, we can see that a new technology, even one that is truly revolutionary at the time of its introduction such as the telephone or electricity, may take a very long time to become fully adopted by a market user.

Yet, we also see that the rate of adoption is speeding up compared to 100 years ago. While it took nearly 100 years for telephones to reach nearly 90% of US households, we see that more recent technological innovations such as the cell phone has achieved almost the same level of market penetration in just over 20 years, less than one-fifth of the time. During this period, society has continued to evolve just as technology has, changing from a largely agrarian-based culture to a very transient postindustrial culture.

These impacts were not lost on the battery and electrification markets. When the first electric vehicles were introduced to the US, neither the average consumer nor retailer had electrical service. So the early adoption by private consumers, which showed some great early progress, was quickly stymied without the means to supply the electricity needed for

The Handbook of Lithium-Ion Battery Pack Design. http://dx.doi.org/10.1016/B978-0-12-801456-1.00002-6

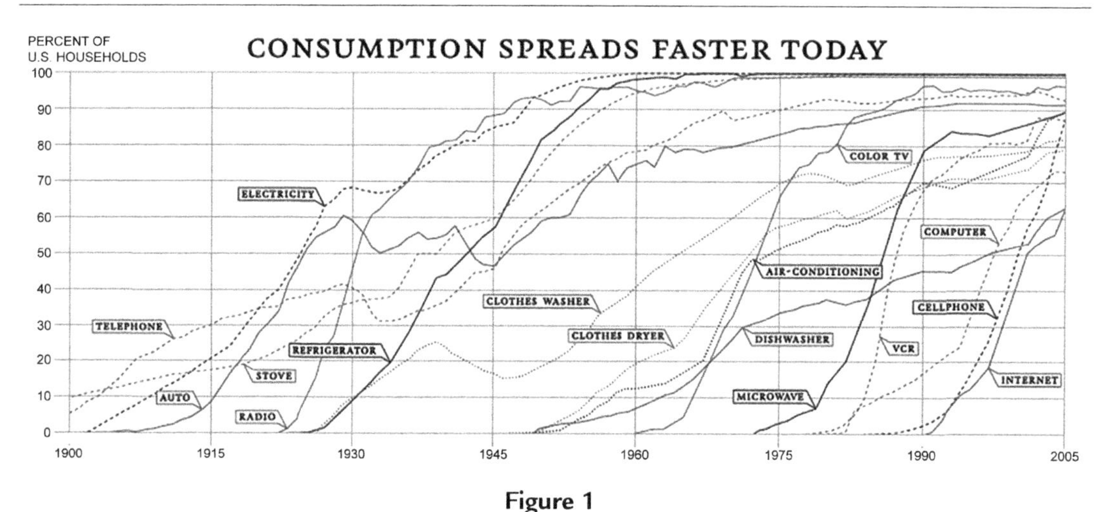

Figure 1
Rate of technology adoption. *New York Times (Felton (2008)).*

charging the vehicles and was in fact limited to either industrial companies or the very wealthy, both of whom were early adopters of electrical service. But even as electrical service availability increased, the electric vehicle demand dropped. There were many potential reasons and causes for this, but they range from the introduction of competing technologies such as the internal combustion engine (ICE) vehicle powered by liquid fuels; the rapid growth of the American interstate system, which increased the distance that vehicles needed to drive; the low cost and great availability of gasoline as a fuel source; and the American's continuous desire for mobility.

So back to the question at hand, will we be forced to repeat the history of electric vehicles and see them die another slow death? Will the market today be stymied again due to the lack of a charging infrastructure? Will range limitations and charging times continue to move consumers away from electric vehicles? But let us back up for just a moment; before we delve into the history of the electric vehicle, we should understand where the enabling technology came from. The introduction and growth of the electric vehicle was only empowered due to the invention of an energy storage technology that offered both high energy storage capability and the ability recharge that energy source—that is where the battery comes into the story. This chapter will attempt to tell the story of how the technology has evolved in order for the reader to gain an understanding of what barriers we may still need to overcome in order to gain true market acceptance and mass market adoption.

The History of the Modern Storage Battery

We generally think of batteries as being a recent invention; however, there is evidence that batteries have been in use for at least the past 2000 years. In the 1930s, a German archeologist was working on a construction site in Baghdad and found something that quite literally

Figure 2
Parthian (Bhagdad) battery.

rewrote the history of the battery. What he discovered during the dig was what looked like a simple clay jar, but inside that clay jar there was a copper cylinder containing an iron rod. The inside of the clay jar showed signs of corrosion that would have been consistent with the jar being used to hold an acidic liquid such as vinegar or wine. This discovery initially proved something of a mystery, but some research and experimentation led to the discovery that this configuration will create a battery capable of generating about 1–2 V of electricity. This discovery has become known as the Baghdad, or the Parthian, battery (Figure 2), based on the era it was believed to have been developed in over 2000 years ago. And while the exact purpose of the battery will very likely forever remain a mystery, it is recognized that Egyptians were known to electroplate fine jewelry around the time this battery would have been in use (Buchmann, 2011) and the battery could have been used for that purpose.

After the Parthian battery was forgotten in the sands of history, there was a very long stretch where little progress was made on battery development. It was until the mid-1700s when two

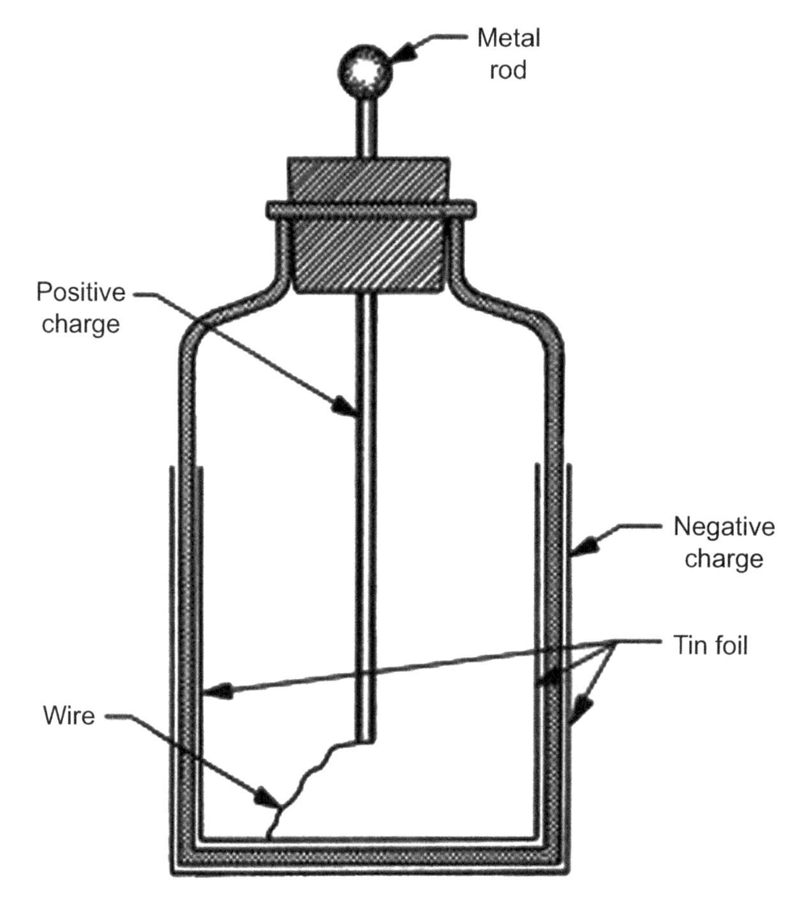

Figure 3
Leyden jar. *HQEW.net (2012). From Webster's New World Dictionary of the American Language: College Edition (1959). All rights reserved. Rights now owned by Wiley. Reproduced here by permission of Wiley Publishing, Inc., Indianapolis, IN.*

inventors, in parallel but separate work, discovered what became known as the "Leyden" jar. Initially this was an attempt to capture the liquid of electricity, yes electricity was believed to be a liquid in many circles of the period, but in the end it proved much more important discovery. The Leyden jar was a device used to store static electricity by lining the interior of a glass jar with a metal foil and adding another layer of metal-foil coating around outside and finally inserting a metal rod through the top and connecting it to the inner foil (Figure 3). The jar was partially filled with water, which acted as a conductor, and when a static generator was applied to the electrode, a static electric charge could be stored in the jar thereby creating the first electrical capacitor (Buchmann, 2011).

It was not until sometime in the mid-1700s that the term "battery" was applied to this type of device and is believed to have been coined by none other than Benjamin Franklin himself, who had a strong fascination with the study of electricity. It is said that Franklin compared a group of

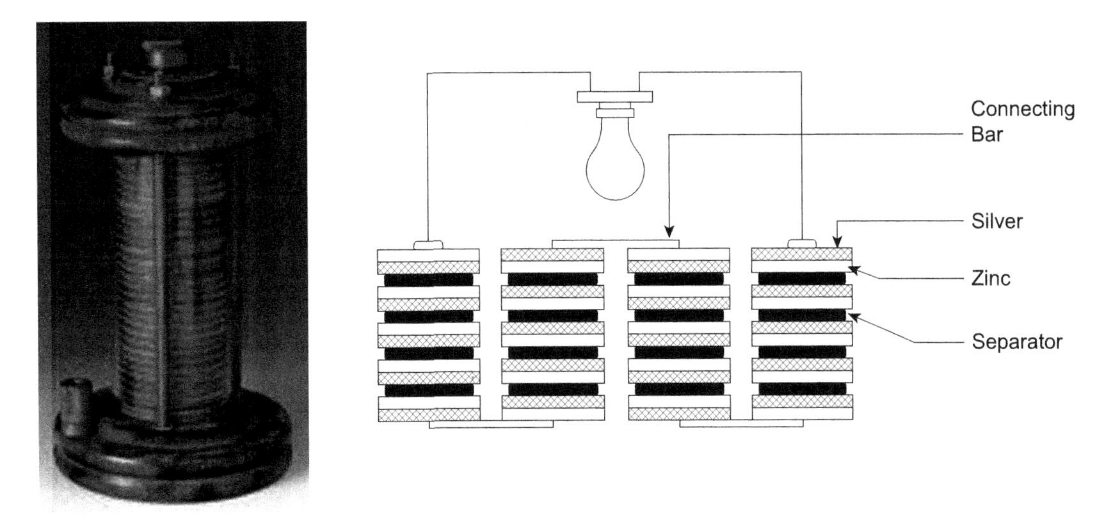

Figure 4
Voltaic pile.

Leyden jars connected together in series with a battery of military cannons thereby coining the phrase "battery" (Franklin & Marshall College, The Phillips Museum of Art, 2008).

Another important discovery in the evolution of the battery was made in Italy in the late-1700s by Alessandro Volta. However, it was not until 1800 when Volta published the results of his years of experiments with what he called a "Voltaic Pile" that his discovery became widely known (Figure 4). Volta's invention proved to be the first electrochemical energy storage cell. The Voltaic pile consisted of two metal plates, one made of zinc and one made of copper, which formed the electrodes, and were separated by a piece of cloth soaked with an electrolyte made of either sulfuric acid or saltwater brine. When the top and bottom contacts were connected by a wire, current would flow with the actual voltage being dependent on the number of pairs of metal plates in the pile (Jonnes, 2004).

The next major evolutionary step in the development of the modern battery occurred in 1859 when Gaston Plante, a French physicist, invented the first rechargeable lead–acid storage battery. Shortly thereafter and based largely on the work of Plante, Camille Faure, another French inventor, improved on that technology by enhancing its current capability in 1881. The work of these two inventors was the basis for the modern lead–acid battery that is used in every automobile today and that sparked the early growth of the both electric mobility and stationary energy storage markets (Buchmann, 2011; Jonnes, 2004).

Not to be outdone, Thomas Edison developed a nickel-based battery in an attempt to compete with the lead–acid battery that was being used in the early electric cars in the early 1900s. But it was not until the early 1970s and into the 1990s before real improvements and innovations in other chemistries were made that really drove the leap in electrification that we see today. But let's hold that thought for a moment while we examine the industrial landscape of the period.

An Electrical Industry Emerges

The fact that battery technology was making major leaps and strides was only possible due to the expansion, growth and availability of electricity, and the electrical networks of the day. Without the "supply" of electricity, the "demand" for energy storage may not have emerged at all. We start looking at the World Fairs of the time because they provide a good representation of the speed of technological innovation and change in the world in general. In the 16-year span between the 1876 "Centennial" World's Fair in Philadelphia, where the only electrical technology shown was an early electric dynamo, and the 1892 Chicago World's Fair that was powered and lit largely by electricity the amount of new and electrically powered technology that was introduced increased significantly. In 1892 Fair, the number of exhibits that focused on electricity or electrical technology grew nearly exponentially during this period and the electric vehicle grew from a pure demonstration to a production product during this same period.

During the 1878 World's Fair in Paris, the first electric vehicles and electric bicycles began to make their appearance in the marketplaces. This World Fair also showcased other major electrical innovations such as Alexander Graham Bell's telephone, electric arc lighting, and even Thomas Edison's phonograph (Gross & Snyder, 2005). And only four years later at the 1892 World's Fair in Chicago, electricity and electrical powered equipment were being showcased in grand style. In fact, the international exhibits hall was powered entirely by Westinghouse's alternating current (AC). The Westinghouse exhibit displayed many new electrical technologies and innovation for the generation and transmission of electricity including an electrical switchboard, a polyphase generator, a step-up transformer, power lines for electrical transmission, a step-down transformer, both a commercial size induction and synchronous motor, and rotary direct current (DC) converters. The Westinghouse exhibit even went so far as to display exhibits of equipment designed by Nikola Tesla himself, including innovations such as a two-phase induction motor and the generators used to power the system. While all of these may seem common place today, remember that this was the first time these had been seen for commercial use. As an interesting side, note General Electric, Tesla and Westinghouse's main competitor had also bid on powering the international exhibit using Edison's DC solution, but ended up submitting a higher cost bid and lost out to the lower priced Westinghouse quote (Barrett, 1894; Jonnes, 2004; Larsen, 2003; Seifer, 1996).

Early Electric Vehicle Development

Between the commercial introduction of the automobile in the late-1800s and the early 1910s, automobiles were powered by one of three different sources: (1) ICEs that were powered by liquid fuel (gasoline); (2) steam engines powered by fire and heat; and (3) electric motors powered by electric batteries using chemical energy. In fact, by 1912, electric vehicles made up over one-third of all US vehicle sales (Kirsch, 2000). During this same period, Europe was also making major advances in electrification not only in personal transportation, such as electric bicycles and automobiles, but also in the area of public transportation, such as trains and mass transit. Yet with all the excitement and growth of electric transportation by the early1920s, the electric vehicle was almost nonexistent except for in a few commercial vehicle and delivery trucks, which limped along until the 1930s or 1940s when even these final few applications disappeared altogether (Kirsch, 2000).

As you can undoubtedly see, as the nineteenth century turned into the twentieth century, electric vehicles seemed poised to become the vehicle of choice for both personal and commercial use. During this period, the prominent Electric Vehicle Association of America (EVAA) pulled together most of the major industry players to focus on driving the growth of the industry. Kirsch (2000) described the EVAA as "...a full-fledged trade organization representing electric vehicle manufacturers, battery makers, and electric companies" (p. 8). However, even with the work of large industry activist groups like this, the growth of the electric vehicle market could not be maintained. Despite having industry-wide support, the electric vehicle was unable to become ingrained in the society and culture of the day.

The first documented and reported sale of an electric automobile in the United States was William Morrison's electric carriage, which was sold to the American Battery Company of Chicago in order to demonstrate their battery technology. In Des Moines, Iowa Morrison developed his six-person, four-horsepower carriage that could achieve a maximum speed of about 14 miles/h (Schiffer, 1994). This, the first produced electric vehicle on the market in the United States, was first demonstrated at the 1892 World's Fair in Chicago, Illinois.

In 1897, a small fleet of electric taxis were put into service in New York City, the first commercial fleet of its kind. This was very typical of the early EV business models during this period. Vehicles, whether they were of internal combustion, electric-, or steam powered, tended to go into service under fleet owners for use in taxis, delivery vehicles, and similar roles.

Those vehicles that were sold to private owners tended to use the historical horse-and-carriage business model. During this period, most of the wealthiest people who used horse and carriages had them housed at a central location where the animals and the carriages were maintained. The early EV market initially attempted to replicate this model using company-hired drivers and housing the vehicles at a central station where they could be recharged and serviced as needed (Kirsch, 2000). This was not only due to the historical horse-and-buggy trend, but as electric power had not yet reached every household it allowed the central power stations to also act as the charging locations for these early electric vehicles.

During this early period, electric vehicles in the United States were being produced by a multitude of companies including companies like Anthony Electric, Baker Motor Vehicle,

Columbia Automobile Company, Anderson (which later became Detroit Electric), Edison, Studebaker Electric, Pope Manufacturing Company, and Riker Electric Vehicle Company (Kirsch, 2000; Schiffer, 1994), just to name a few. And the EV offered many benefits to the early consumer. Compared to the early ICEs, it was quiet, generated no harsh vibrations, and no smell. It also did not require the complex manual shifting that was necessary in the ICEs of the time as the EV used a single electric motor. The EV also did not require the manual starting of the ICE, which was a major advantage to female drivers. By the early 1910s, most homes in large US cities were being wired for electricity. In the early cities of the 1800s and 1900s, the short driving range was also not a major challenge to gaining early market acceptance. All of these factors helped to push the electric vehicle to a high level of early market acceptance.

However, there were also some major challenges with both the technology and infrastructure of the time, largely due to technological limitations of the day. For instance, the early DC motors limited the speed of the vehicles to about 20 miles/h. From an infrastructure standpoint, electricity was being installed in many private homes about this time but a vast majority of the areas were without either private or public power services. But even with the technologies of the day, many of the EVs were able to achieve electric drive ranges between 50 and 100 miles—not too much different than those of today. As the price of gasoline dropped and it became very widely available, range began to become an issue. Additionally, ICE automotive technology was improving at a rapid pace. The introduction of Henry Ford's low-priced Model T in 1908 and the introduction of the automatic starter by Charles Kettering in 1911 both had significant impacts on the electric vehicle. Finally, the last coffin nail of the EV was firmly pounded into place as the US highway system expansion began connecting American cities.

These were several of the main factors that caused the electric vehicles to become obsolete very early in their life. While several models continued to be produced into the 1940s by the 1950s, there were no longer any electric vehicles in production in the United States.

Modern Vehicle Electrification

As the oil embargo of the 1970s began to make companies again look for more fuel-efficient options, most of the major automakers began experimenting with various types of electrification, turbines, and a variety of different fuel types. During the 1980s, the major automakers continued doing research into vehicle electrification but no products made it to market during either of these periods.

It was really not until we entered into the 1990s that the major automakers work on hybrid and electric vehicle solutions began to produce results. In parallel with these advancements, the first commercial lithium-ion battery cells were introduced to the market in 1991 and were quickly followed by the modern nickel-metal hydride (NiMh), which was introduced in the mid-1990s. With the rapid spread of personal electronics, these high energy-density batteries became the energy storage solution of choice for many different applications from portable electronics to hybrid and electric vehicles.

One of the biggest drivers in the United States for this new effort to electrify vehicles occurred in 2000 when California passed a Zero Emissions Vehicle (ZEV) mandate. This was one of the first pieces of legislation in the United States that required manufacturers to offer vehicles that emitted no emissions whatsoever; essentially this legislation mandated the introduction of electric and fuel cell vehicles. This was actually an evolution of an earlier ZEV legislation passed in 1990 as part of California's Low Emission Vehicle (LEV) program that was intended to begin stemming the growing amount of air pollution in California.

Perhaps, the first truly production-intent electric vehicle was General Motors' (GM) EV1. It was intended to be an optimized vehicle platform designed from the ground up entirely for electrification. The earliest versions of this vehicle used lead–acid (PbA) batteries, but later vehicles switched to nickel-metal hydride (NiMH) by Ovonic Battery Company, which reduced the size of the battery by about 50% while offering about the same amount of energy storage capacity. And while only 1117 vehicles were produced during its production run between 1996 and 1999, this vehicle became the basis for many of the electric vehicles that followed and formed a cult following among enthusiasts and early adopters.

General Motors followed the EV1 with the development of a "Parallel Hybrid Truck" (PHT), and then joined in a partnership with Chrysler, Daimler, and BMW to develop a 2-Mode Hybrid System. The 2-Mode was a strong hybrid that integrated a 300-V NiMH battery under the rear seat. At the same time as the 2-Mode was in development, GM also developed a "Belt-Alternator-Starter" (BAS)-type mild hybrid system. The first introduction of GM's BAS was a 36-V system with a NiMH battery developed by Cobasys (Cobasys, 2006); however, the second generation (now referred to as e-Assist) increased the voltage of the system to 115 V and changed to a 0.5 kWh lithium-ion air-cooled battery designed by Hitachi Vehicle Energy Ltd (Dlegs Hybrid Cars, 2011). All of these were the predecessors to the development of GM's Voltec technology, which debuted in the Chevrolet Volt. The Volt is a "series hybrid" that combines both a small ICE with a 355-V lithium-ion battery with cells from LG Chem and pack designed by GM and two electric motors.

During the same period, Toyota began working on a system to hybridize their small cars. Similar to GM efforts with the EV1, Toyota designed a hybrid vehicle from the ground up. The Toyota Hybrid System (THS) they developed formed the basis for what has become the best-selling hybrid vehicle on the market today, the Toyota Prius. Toyota's hybrid system used an air-cooled 288-V NiMh battery with about 1.7 kWh of energy as the basis of the energy storage portion of the THS. Part of the strong market acceptance of the Prius was due to its unique look and aerodynamic design, it became an "image" vehicle for people trying to make an environmental or political statement and very quickly becoming the standard for high fuel-economy hybrid systems. After over 10 years of success in the marketplace, Toyota expanded the THS lineup by offering the THS in the Lexus brand, making Prius a stand-alone brand by adding the Prius C, a small city car, the Prius V, slightly larger hatchback, and a Plug-In Hybrid (PHEV) with about a 10-mile range. Toyota also developed a fully electric RAV4 SUV that it offered for lease between 1997 and 2003 and later partnered with Telsa to develop the second generation electric RAV4. The second generation RAV4 EV battery was based on the Tesla Model-S battery pack and was a 386-V lithium-ion battery with about 52 kWh of total energy onboard.

Honda followed with the introduction of the Honda Insight, a hybrid based on their integrated motor assist (IMA) technology, which actually *preceded* the Toyota Prius and was the first produced hybrid vehicle available in the US market. Like the other OEMs, the Insight was also a ground-up design that had a major focus on low weight and aerodynamics, however it had a fatal flaw—it was designed as a "two-seater." And while there were a number of enthusiasts and early adopters, the two seat configuration limited the market potential and ended up with global sales below 18,000 total units between 1999 and 2006 when it ended its production run. One final note on the Insight, as of this writing in 2014 is that the Insight is still the highest fuel-economy gasoline-fueled hybrid vehicle as certified by the Environmental Protection Agency in the United States. Honda followed that with by adding their hybrid technology to their best-selling sedan, the Honda Civic. The Civic quickly became the second best-selling hybrid in the world, second only to the Prius. The Civic used a 144-V NiMh battery with about 0.8 kWh of energy—about half the size of the Prius battery. Honda also offered a CRZ hybrid with a lithium-ion battery in 2010 and a fully electric Fit with a 20 kWh lithium-ion battery in 2013.

Not to be left behind, the Japanese and Korean auto manufacturers began offering hybrid and electric vehicle options. Mitsubishi introduced the fully electric i-Miev; Mazda offered hybrid options on their Tribute, Mazda3, and Mazda6; Hyundai a hybrid Sonata, Tuscon, and Elantra; Kia with a hybrid Optima; and Subaru with an XV Crosstrek and a Stella Plug-in Hybrid.

During the late-1990s and early 2000s, Nissan took a bit of a "wait-and-see" attitude toward hybrids and electric vehicles. However, once they made a decision to develop a fully electric vehicle, they created a large world-class development team with a single focus—to develop an affordable electric vehicle. The result was the Nissan Leaf; as of 2014, it was the best-selling mass production electric vehicle in the world. Renault–Nissan also partnered with the Battery Swapping start-up Better-Place in France for a short period of time to offer vehicles with the ability to "swap" a depleted battery for a fully charged one. Renault built on the success of the Leaf with the introduction of the Fluence, Kangoo, and Twizzy, all are fully electric vehicles based on the Nissan Leaf technology. Nissan's focus on fully electric vehicles meant that their focus on hybrid vehicles was somewhat delayed with the Nissan Altima being introduced in 2007, which licensed the THS. This move allowed Renault–Nissan to maintain their focus on electric vehicles, while still allowing them to offer hybrid solutions.

In the United States, Ford and Chrysler were somewhat late followers to the hybrid and electric vehicle market. Chrysler, through the Daimler–Chrysler joint venture, used the jointly developed 2-Mode hybrid for several experimental and research fleets and developed a PHEV Ram pickup truck for a U.S. Department of Energy (DOE) program. Chrysler continues to lag in their introduction of production hybrid and electric vehicles but have several new programs under development so expect to see announcements in the near future. Ford, on the other hand, followed a similar strategy as Nissan and bought themselves some time by licensing the THS to launch in the Ford Fusion. This allowed Ford time to develop their own system, which continued to make the Ford Fusion hybrid a very strong seller in the United States. Ford followed this with a PHEV model, the C-Max. Ford also utilized the services of automotive Tier 1, Magna and their E-Car division to develop the fully electric Ford Focus. The Focus EV uses a 23 kWh lithium-ion battery with cells provided by LG Chem and the pack itself done by Ford.

With continuously falling CO_2 emissions targets in Europe with the enactment of Euro 5 and Euro 6 emission standards, the European automakers have also been working hard on introducing new production hybrid and electric vehicle solutions. BMW, Daimler, Audi, Fiat, Peugeot, and VW, among many others, are all working on both hybrid and fully electric vehicles introducing vehicles such as the Smart Fortwo, Fiat 500e, and the Think!. BMW now offers a full portfolio of hybrid, plug-in hybrid, and electric options on vehicles such as the e-Tron, the i-8, and ActiveHybrid by combining luxury and performance with fuel economy under the "Efficient Dynamics" moniker. VW introduced several entries into the hybrid, plug-in, and electric space including the e-Golf, the Audi A3 e-tron Plug-in hybrid, and Golf GTE.

In China, automakers began working on electrification solutions in the early 2000s due to the rapid growth of personal automobiles and the air pollution that this growth has caused, especially in the large cities, which drove governmental incentives and regulatory emissions requirements. Companies such as BYD, which started out as a lithium-ion cell manufacturer for small consumer electronics, grew to become a full-vehicle manufacturer offering hybrid, plug-in, and fully electric vehicle options as an outlet for their cell production. Other Chinese manufacturers followed suit, including Beijing Automotive Industry Corporation (BAIC),

Table 1: Levels of automotive electrification.

	Micro Stop/ Start	Mild-HEV	Full-HEV	Plug-in Hybrid	Plug-in E-REV	H_2 Fuel Cell EV	Electric Vehicle
Functionality	Engine stop-start at idle	Engine off during deceleration, mild regenerative braking, electric power assist	Full regenerative braking, engine cycle optimization, electric launch, limited pure electric drive, allows engine downsizing	Grid rechargeable, extended electric drive during charge depletion mode, high fuel economy during short trips	Full function electric drive, initial pure electric range, significantly reduced gas, refueling, zero fuel on short trips	Full function electric drive, petroleum free, emissions free	Plug-in recharge only, 100% pure electric range, no refueling
Battery type	Power	Power	Power	Power/energy	Energy/power	Power/energy	Energy
Battery chemistry	PbA, lithium-ion	NiCd, NiMH, lithium-ion	NiCd, NiMH, lithium-ion	Lithium-ion	Lithium-ion	Lithium-ion	Lithium-ion
Battery pack size	250–1000 Wh	1–1.5 kWh	1.5–3 kWh	7–15 kWh	15 kWh+	TBD	15 kWh+
Electric range	None	None	<1 mile	10–30 miles	35+ miles	300+ miles	75+ miles

Geely, Shanghai Automotive Industry Corporation (SAIC) initially through their joint venture with GM, Chang'an, Chery, Dongfeng, First Auto Works (FAW), Brilliance Automotive, Foton, Great Wall, Lifan, and many others.

In addition to the major OEMs, the 2000s brought the introduction of many smaller manufacturers offering new EV options to the marketplace. Some of them, such as Tesla, were very successful after some early challenges. While others, such as Fisker, Azure Dynamics, and CODA Automotive, ended up filing bankruptcy, they did make significant inroads for electric vehicles.

This is clearly not a complete listing of all hybrid, plug-in hybrid, and electric vehicle offerings, but is intended to give you an idea of the scope of change and the amount of options that have come into the market over the last decade. It also shows us how different the marketplace is from a century ago when the first electric vehicles were introduced to the general public. Today these vehicles are common place and are quickly becoming the "standard" and reaching mass market acceptance, at least for the hybrids, rather than simply science experiments for early adopters.

Today, electric and hybrid vehicles are clearly here to stay. By the early 2020s, I would expect to see almost every vehicle having at least a very simple Stop/Start type microhybrid as its base offering. Will fully electric vehicles become "mainstream"? That is much more difficult to prognosticate; however, I think that as technology continues to improve the option for fully electric vehicles will become better options for many drivers.

In the meantime, I believe that PHEVs are a great "transitional" technology for much of the population since they bridge the gap between standard ICEs and advanced fully electric vehicles. The PHEV vehicles eliminate the concern over "range anxiety" as they continue to offer the same, or greater, driving range as traditional ICE's, while still offer between 10 and 40+ miles of electric driving range. This will enable the general consumer time to get comfortable with the new technology.

Table 1 breaks out each of the general categories of hybridization from the simple Start/Stop all the way to fully electric vehicle and highlights the differences in these different options.

In the next chapter, we will begin reviewing all of the different terminology that is used for both the electric vehicle as well as for the advanced batteries in order for us to be able to "speak the same language." As with any other industry, the advanced battery industry has generated quite a unique set of acronyms and terminology that can get quite confusing if you are not familiar with them. For this reason, Chapter 3 is a very good starting point for the rest of the book.

CHAPTER 3

Basic Terminology

Perhaps even more confusing than figuring out the size of a battery system or what type of battery is needed is understanding the terminology surrounding it. This is because many of the acronyms tend to have multiple definitions and meanings. One good source for understanding many of the different terms, and one that is far more complete than this brief summary, is the specification from the Society of Automotive Engineers (SAE) J1715 (Society for Automotive Engineers, 2014). However, remember that this list is specific to automotive applications; other applications in portable power, grid, and stationary energy may use different terminology.

Vehicle and Industry Terms

To begin simply, the abbreviation EV generally refers to an Electric Vehicle (EV), a vehicle that is fully electrified with no internal combustion engine (ICE) whatsoever. All of the power that is provided is done through the electric battery, powering one or more electric motors to provide propulsion and power to all of the other systems on the vehicle. These are also sometimes referred to as battery electric vehicles (BEVs), both terms are used interchangeably. However, the same abbreviation EV is also used to mean *Electrified* Vehicles (EVs), which refers to vehicles with all forms of electric power support including microhybrids, hybrid electric vehicles (HEVs), plug-in hybrid electric vehicles (PHEVs), and battery electric vehicles (BEVs). For the purposes of this book, I will use EV to mean only electric vehicles.

A PHEV is a vehicle that has both an electric propulsion system, generally ranging from 10 to about 40 miles of pure electric drive range, along with a combustion engine. This type of hybrid is often a parallel hybrid. In simplest terms, this means that the engine works in parallel with the electric motor. In this configuration, the electric motor is often "sandwiched" between the engine and the transmission. The benefit of the PHEV configuration is that it offers the full driving range that a comparable ICE vehicle would have—usually between 350 and 500 miles depending on the vehicle (Figure 1).

Similarly, the Extended Range Electric Vehicle (EREV), also referred to as a Range Extended Electric Vehicle (REEV) or sometimes as a Range Extender (REX), is considered a "series hybrid." In a series hybrid configuration, the electric motor(s) is(are) generally operated instead of the engine and the motors are not "in-line" with the engine, so you can use one or the other and sometimes both at the same time. In this configuration, the electric motors always power the vehicles propulsion; however, the power source switches between the

The Handbook of Lithium-Ion Battery Pack Design. http://dx.doi.org/10.1016/B978-0-12-801456-1.00003-8

electric battery and the ICE with the engine getting engaged and operating like a generator once the battery has reached a predefined minimum state of charge and keeping the battery voltage maintained at that preset level until the vehicle is recharged from the grid (Figure 2).

A common misconception of the PHEV and series hybrid configuration is that the engine will recharge the battery in this mode. This is not actually how they operate in the current designs. The engine will operate as a generator to power the electric motors but will only maintain the battery at its minimum state of charge until the vehicle can be plugged into the grid to recharge.

The term HEV refers to the most common configuration of electrified vehicles. These can range from the mild hybrid to a strong hybrid. A mild hybrid will have a smaller battery, usually less than one kilowatt hour (kWh), and will provide less electric power to the system, whereas a strong hybrid will have a slightly larger battery, often about 1.5 kWh, and will provide some minimal amount of electric propulsion in addition to powering some of the auxiliary systems. The term hybrid comes from the act of hybridizing of an ICE with a battery electric powered system, essentially creating a dual power system. In the hybrid car, the electric motor will generally provide power support throughout the operating cycle, but does not provide electric driving capability (mild hybrid) or at best only minimum electric drive capability (strong hybrid). During acceleration (Figure 3), the battery power will be added to that of the engine power to reduce the overall load on the engine, thereby improving fuel economy and reducing emissions. During deceleration, the motor recharges, or

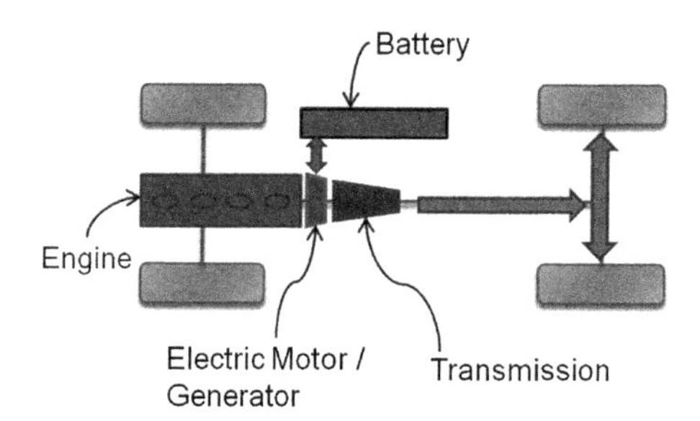

Figure 1
Parallel hybrid configuration.

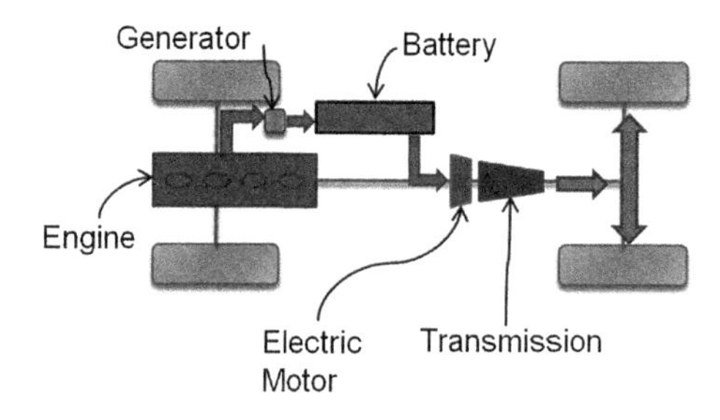

Figure 2
Series hybrid configuration.

regenerates energy, back into the battery, which then provides power to the vehicle during start and stop events when the engine is turned off. It is important to understand that in the hybrid system the electric power to charge the battery is entirely provided by the on-board generator, the HEV does not plug into the grid to charge the battery (Figure 3).

A microhybrid (μ-HEV) refers to a Stop/Start (S/S)-type system. In this type of hybrid, there is a very small battery, often a lead acid or a very small lithium-ion battery, that powers the vehicle systems when the control system automatically shuts off and then restarts the engine when the vehicle stops. Most of these systems do not offer support for regenerative braking because the battery is too small to capture the amount of energy that would be regenerated back into it. However, some systems with larger batteries (usually those with >750–1000 Wh) may include regenerative braking capability and on-board power support for electrically powered systems such as power steering, power brakes, entertainment system, HVAC, and vehicle lighting while the engine is stopped. This configuration is gaining much popularity in all regions around the world due to the low cost and general simplicity of this system.

A Neighborhood Electric Vehicle (NEV) is a fully electric vehicle but it is limited to a maximum speed of about 30 mph (up to 45 mph in some areas), must have a gross vehicle weight of less than 3000 pounds, and can be driven only on limited public streets. These limits are based on current US laws and regulations for this type of low-speed vehicle. The NEV is commonly used on college campuses and in small private communities. These vehicles can basically be considered next generation or "souped up" golf carts as that is their basic beginning structure. One important thing to take note of is that this same term has a very different meaning in different regions of the world. In China, this same term, NEV, is used to describe China's New Energy Vehicle (NEV) policy. This policy was develop in China based on the "863" program and subsequently the Tenth Five Year Plan, which created a set of incentives and goals for increasing the amount of New Technology Vehicles in an effort to reduce air pollution. This policy was also developed in order to advance Research and Development (R&D) in the areas of HEVs, EVs, and Fuel Cell Vehicles (FCVs) in China. For the purposes of this text the term NEV will be used to represent the Neighborhood Electric Vehicle unless otherwise stated.

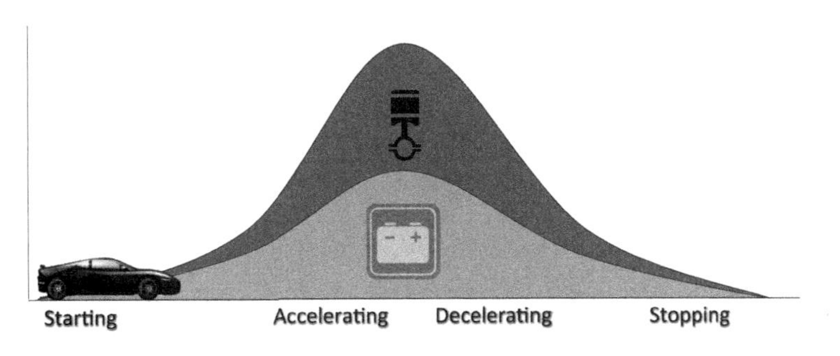

Figure 3
Operating cycle of hybrid car.

There is another term that is occasionally used in reference to electric vehicles, that is Light Electric Vehicle (LEV). The LEV ranges from both two- and three-wheeled electric motorcycle-type platforms to electric bicycles. The LEV market is actually the largest electric vehicle market in the world with between 25 and 30 million LEVs sold annually throughout the world. The vast majority of these are electric bicycles (e-bikes), pedelecs (electric assist bicycles), and electric scooters that are sold in Asia, but demand is growing strongly in the Western markets for this type of technology.

There are also a couple of EV industry terms that should be understood. The California Air Resource Board (CARB) or sometimes referred to as the Air Resource Board (ARB) is one of the biggest reasons for the growing demand for electric vehicles in the United States. In 1990, CARB passed into legislation the Zero Emissions Vehicle (ZEV) mandate, which required that by 1998 two percent of all vehicles sold in California must emit zero emissions, increasing to 5% in 2001 and 10% in 2003 (California Air Resource Board, 2014). While this early mandate for vehicle electrification missed its initial targets, it has become the driving force for both electrification and hybridization in the United States. By 2009, there were at least 21 other states that had either adopted ZEV mandates or were considering doing so. In 2013, eight US states including California, Connecticut, Maryland, Massachusetts, New York, Oregon, Rhode Island, and Vermont signed a memorandum of understanding (MOU) to put 3.3 million ZEVs on the roads by 2025 through a combination of R&D, infrastructure, education, and incentivization. This move would increase ZEV penetration to about 15% of all vehicle sales in the United States (Green Car Congress, 2013) (Figure 4).

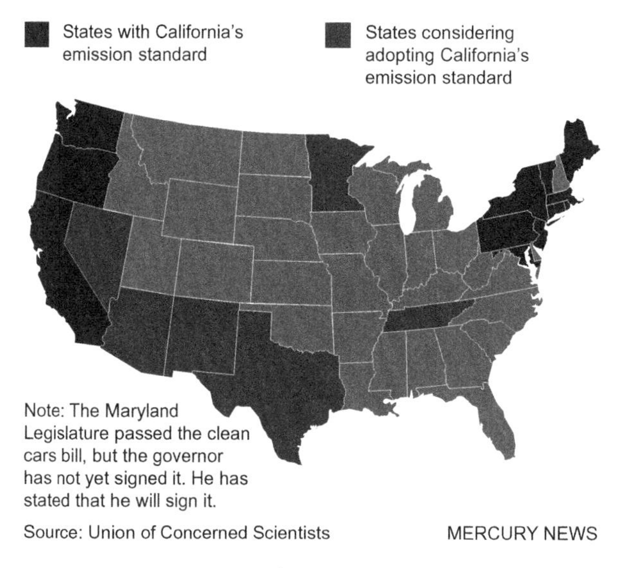

Figure 4
States that have enacted ZEV emissions standards.

The other legislative reason automakers began looking at a portfolio of solutions to improve fuel economy was the US Corporate Average Fuel Economy (CAFE) standards that were enacted in 1975. These standards were originally put in place as a result of the Organization of the Petroleum Exporting Countries (OPEC) oil embargo of the early 1970s in an effort to reduce the US reliance on foreign oil. Initially the law required that automakers passenger fleets achieve 18 miles per gallon (MPG) by 1978, increasing to 27.5 mpg by 1985. However, during the Regan administration, the law was relaxed down to 26 mpg and then increased again to 27.5 mpg by 1990 and then for over 21 years the CAFE laws remained fixed at 27.5 mpg. In 2006, the CAFE laws were reformed and a new manner of measuring CAFE was installed in 2011, by vehicle footprint instead of weight. And a new set of fuel economy targets were put in place between 2011 and 2015 raising the target from 27.5 mpg up to 35.5 mpg by 2015 (2016 model year). In 2012, the next generation of standards were agreed to that define the required improvements between the 2017 model year and 2025 model year, resulting in a fleet average fuel economy in 2025 of 54.5 mpg (National Highway Traffic Safety Administration, 2014).

Stationary and Grid Terminology

We will also briefly describe several of the key terms that are used in the large stationary energy storage market before we delve into the battery specific terminology. Stationary and grid-based energy storage generally refer to the large, often megawatt-sized systems that utility companies put in place in order to provide several functions to support the grid operation. These functions include providing storage for intermittent and renewable generation sources such as wind and solar (renewables), providing fast demand response, load leveling, managing the power quality, and ensuring reliability of their power generation services.

The term Community Energy Storage (CES) refers to a relatively small energy storage device that often ranges from about 25 kWh up to about 100 kWh in size. This technology is intended to be used by utility companies to install capacity in specific neighborhoods where they see increased demand and/or that are at the periphery of the grid. This enables both "islanding," the ability to operate a microgrid independent of the larger grid, as well as support during downed power events. One organization, the Energy Storage Association (ESA), has published an "open-source" set of requirements for these types of application in an effort to begin industry standardization and communization (Energy Storage Association, 2014).

The CES falls into the category of energy storage systems (ESSs) referred to as Distributed Energy Storage (DES) systems. In general, the DES refers to systems much larger than the CES ranging from hundreds of kilowatt hours up to megawatt hours (MWh) of installed capacity. The DES systems are generally installed at utility-owned location such as the electrical substations and offer some of the same benefits as the CES systems, only at a much greater scale.

Battery Terms

Before we begin talking about battery design, we need to first understand some of the basic terminology surrounding the battery. Please note that this list is not intended to be an exhaustive list but rather is an explanation of some of the most common terms that you will find in the lithium-ion battery industry. For a more exhaustive list of terms and definitions, again I would refer the reader to one of the industry standards organizations such as the SAE standard J1715 *"Battery Terminology"* (SAE International, 2014).

Ampere—Often referred to as "Amp," this is a unit of measurement of the battery current.

Anode—The anode is the negative "–" terminal inside the battery cell. It is typically a thin piece of highly conductive aluminum or copper that is coated with graphite, carbon, or other similarly conductive material.

Battery Management System (BMS)—The BMS is the control system within the battery pack that consists of one or more electronic controllers, which manages the charging and discharging, monitors the temperature and voltage, communicates with the vehicle system, balances the cells, and manages the safety functionality of the battery pack.

Beginning of Life (BOL)—The term BOL refers to the battery energy, capacity, and power when it is first built or is at the beginning of its life.

C-rate—The term C-rate is important; it refers to the rate at which a battery can charge or discharge all of its energy (or power). In other words, it describes how fast a battery can accept a charge or give up its power (discharge). C-rate is described in relation to a 1 h discharge, so 1C-rate is equal to the rate at which a battery is fully discharged (or charged) in 1 h. Along the same lines, 2C-rate would then be equal to the rate at which a battery is fully discharged in 30 min (60 min/2C = 30 min). If the C-rate number goes up the discharge time goes down, and vice versa. So 0.5C discharge rate would be equal to a 2 h discharge period (60 min/0.5C = 120 min or 2 h).

Capacity—Capacity is measured in Amperes (A) and is a measure of *amount* of energy in a system. Think of capacity as being analogous to the size of the hose that water may flow through. Larger capacity is equal to a larger hose. If we add in the aspect of time, Ampere hours (Ah) is akin to Miles per hour (mph).

CARB—The CARB is a legislative group in California with the mandate of passing legislation that will improve the air quality of the state. CARB is often also referred to "ARB" for the Air Resource Board.

Cathode—The cathode is the opposite of the anode; it is the positive "+" terminal inside the battery cell. It is typically a thin piece of aluminum or copper that is coated with lithium-ion chemistry such as lithium-iron phosphate (LFP), lithium cobalt oxide (LCO), lithium-nickel/manganese/cobalt (NMC), lithium-manganese oxide (LMO), or other lithium-based chemistry.

Current—Current is the measurement of the flow of electrical charge, which may be carried by electrons moving through a wire or circuit board and by ions moving through an electrolyte between the anode and cathode and is measured in Amperes.

Cycle—The term cycle refers to the process of discharging and then charging a battery. A complete discharge and then charge is known as one cycle. A cycle may run at various levels of power and/or voltage or even using a constant rate of charge and discharge, depending on the application requirements. A cycle may be full, completely discharging and then charging the cell, or partial, only discharging to a set level and then recharging back up to starting level.

Depth of Discharge (DOD)—The DOD is a measurement of how much of the cell or pack energy will be used for that application. You will typically use lithium-ion battery only somewhere between 20% and 90% of the total amount of energy in order to prevent overcharge at the top and manage low end voltage. So if we think of the battery as a 10-gallon gas tank PHEV, it may only use 60%, or 6 gallons, of its energy—so the DOD would be 80%. For a HEV or micro-HEV, the battery may only use 30–50% of the available energy or 3–5 gallons in this example. And an EV may use 80–90% of the total energy available, about 8–9 gallons in this example. During system development one design decision that is commonly made, especially for hybrid applications, is to reduce the Depth of Discharge (DOD) during operation in order to achieve greater cycle life and improve safety in the energy storage system.

Electrodes—The term electrodes refers to the combined pair of anode and cathode when it is assembled in a battery cell.

Electrolyte—The electrolyte is liquid, gel, or other material that is used as the medium to transfer the lithium-ions back and forth between the anode and cathode.

Electric Miles per Gallon (eMPG)—eMPG refers the "electric miles per gallon" equivalent rating that is used by the US Environmental Protection Agency (EPA) to offer an equivalent comparison for consumers to traditional mpg used on ICE vehicles.

End of Life (EOL)—The EOL of a battery is reached when the battery's maximum power and energy have been reduced to about 80% of their BOL measurements. The general rule of 80% is based on either power or energy that has dropped to a point where acceleration (power) or range (energy) is no longer "satisfactory" to the consumer. However, depending on the application, a lower EOL could be used.

Energy—The term energy, which is measured in kilowatt hours (kWh), refers to the amount of energy that a battery will store, think of it as being analogous to the size of the gas tank.

Energy Density—Energy density is the measurement of how much energy a cell or pack contains in relation to either its mass or its volume. Energy density is measured either in watt hour per kilogram (Wh/kg) or as Watt hour per Liter (Wh/L). When it is used to compare to mass in the Wh/kg, it is referred to as Gravimetric Energy Density. When it is referred to in relation to volume in the Wh/L form, it is referred to as Volumetric Energy Density.

Energy Storage System—The term ESS is used in various forms, but it generally means the complete battery pack system. The ESS is the combination of cells that are mechanically and electrically connected, along with the appropriate thermal, electronics, and mechanical structure to house the entire unit. In essence, it is everything that is "in the battery box."
High Voltage (HV)—any system with a voltage over 60V is considered "high voltage" and must include appropriate protections (HVIL, safety disconnect, orange cabling, etc.) in order to prevent harm to the workers or anyone who may come in contact with the system.
Impedance—Impedance is basically a measurement of how much the materials within the cell slow down, or impede, the flow of current when a voltage is applied. It is measured in Ohms and is represented by the symbol Ω. AC impedance is generally used in relation to measuring at the cell level. Impedance basically builds on the concept of DC resistance, which is a measure of magnitude (where there is no difference in phase shift between voltage and current), by adding the concept of phase to the measurement.
Jelly Roll—The jelly roll is the combined assembly of anode, separator, and cathode that are either stacked or rolled together and inserted into the can or pouch (Figure 5).

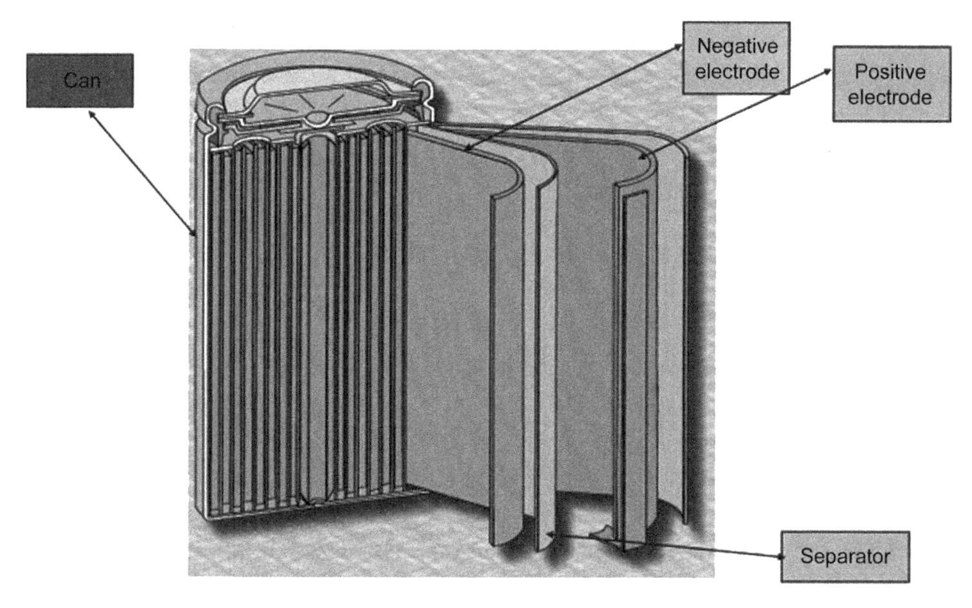

Figure 5
Schematic of a cylindrical lithium-ion battery.

LIB—An acronym that is occasionally used in place of the terminology lithium-ion battery or li-ion.
Miles per Gallon—This is the standard measurement of fuel efficiency that is applied to almost all ICE vehicles in the United States. In other regions, kilometers per gram CO_2 (kg/CO_2) is used for the same purpose.

Parallel—A battery with parallel connections refers to cells that are connected in parallel (e.g., positive to positive, negative to negative, etc.). In a parallel connection, you are feeding current into all of the cells at the same time and pulling current out of them at the same time. When connecting cells in parallel, the system capacity is increased. The example below represents three cells: let us assume that they are 3.6V and 5Ah; in a parallel configuration, this would end up with still at 3.6V but the capacity would increase to 15Ah (5Ah×3 cells) (Figure 6).

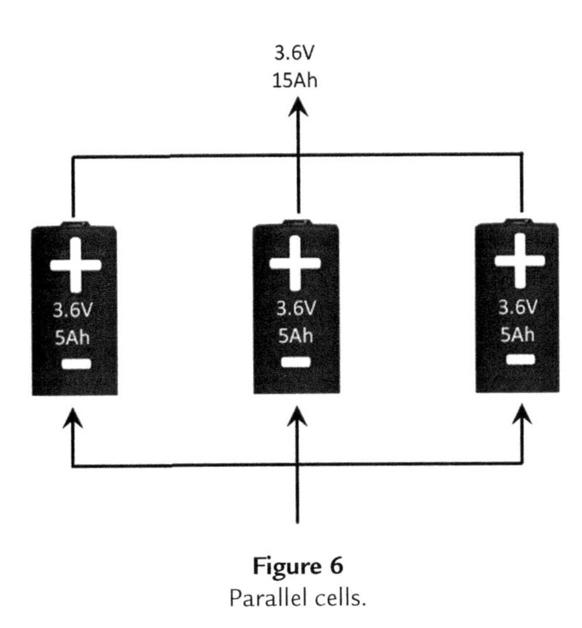

Figure 6
Parallel cells.

Power Density—Measured in either kilowatt per kilogram (kW/kg) or kilowatt per Liter (kW/L). Similar to energy density, power density is the comparison of a battery's power in relation to its weight or volume.

Power Net—The term power net is a relatively new one. It is getting used largely in European countries by automakers to refer to the collection of on-board technologies and software that is being run electronically. It usually includes the infotainment systems, radio and communications systems, navigation systems, and anything else that is electrically powered.

Primary Battery—A primary battery is simply a battery that is not rechargeable. An example of a primary battery would be the AA-, C-, or D-type battery, which you might use in your household electronics that are often made using an alkaline chemistry.

Resistance—Resistance is related to impedance and is a measurement of how much the materials within the cell slow down, or resist, the flow of current when a voltage is applied. It is measured in Ohms and is represented by the symbol Ω. DC resistance is generally used in relation to measuring at the pack level. DC resistance describes the magnitude of the difference between voltage and current (assumes no difference in phase shift between voltage and current).

Secondary Battery—A secondary battery is simply a battery that is rechargeable; examples include lithium-ion, nickel-metal hydride, and lead acid.

Separator—The separator is a thin sheet of material, often a single- or multilayer plastic (polypropylene) or ceramic based material, that separates the anode from the cathode in order to prevent them from touching and creating a short circuit. The separator also must allow the lithium-ions to pass between anode and cathode.

Series—A series configuration refers to a grouping of cells connected in series (e.g., negative to positive, etc.). Connecting cells in series increases the voltage of the overall system. The example below represents three cells: let us assume that they are 3.6 V and 5 Ah similar to our previous example, in a series configuration. This configuration would end up with 10.8 V (3.6V × 3 cells) but will remain at 5 Ah. Putting lithium-ion cells in series is similar to connecting multiple garden hoses end to end, just as the garden hoses are in series so are the batteries. The difference is that in batteries connecting them in series increases the voltage but not the capacity—the same effect does not occur when connecting garden hoses in series (Figure 7).

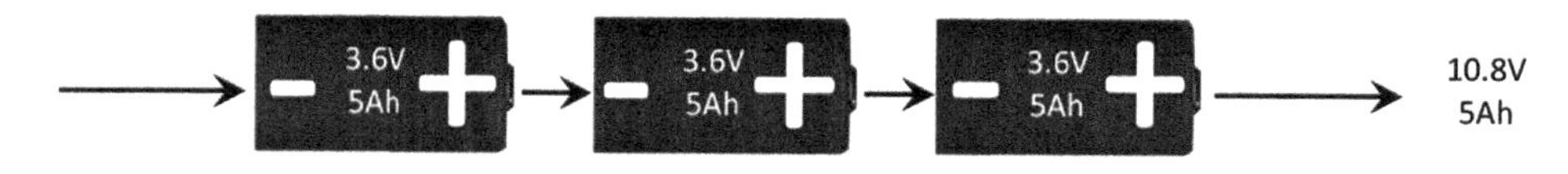

Figure 7
Lithium-ion cell in series connection.

Short Circuit—A short circuit occurs when the positive and the negative poles or electrodes of a battery are connected. In essence, a short circuit creates a circular connection within a cell, driving all of the current back into the cell or pack which will eventually, and usually very quickly, lead to a catastrophic failure. This could occur inside the cell due to the growth of dendritic materials that grow between the anode and cathode connecting them electrically. This could also happen if a microscopic piece of debris gets assembled in the jelly roll, which could eventually pierce the separator and connect the two electrodes. If it happens inside the cell, it is an "internal" short; but if the electrical connection is made between the poles outside of the cell, it is referred to as an "external" short (e.g., outside of the cell).

State of Charge (SOC)—While DOD measures how much of the battery is being used, SOC measures how much is left at a specific point in time. SOC is usually measured from 0% to 100%, but it is measuring against the DOD. Basically the SOC tells you how much energy or power you have right now, it is like the gas gauge in your car (Figure 8).

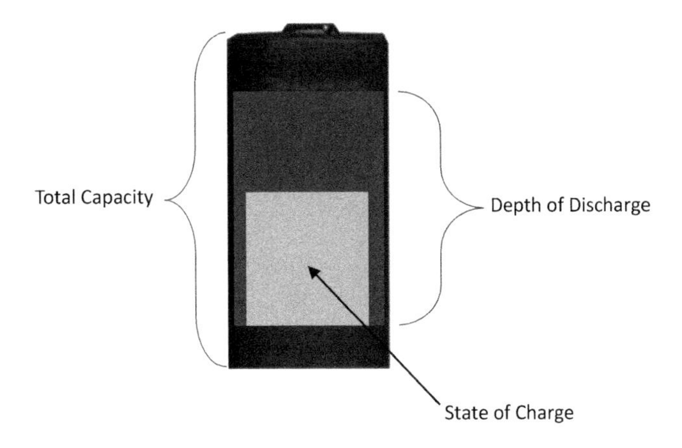

Figure 8
DOD, SOC, and total capacity of a lithium-ion cell.

State of Health (SOH)—SOH is an interesting measure because it does not represent any "standard" measurement; different battery companies, controls companies, and application users may have different definitions. In general, SOH refers to the current state of health of the battery as compared to its beginning of life measurement. In other words, SOH is intended to tell you how long the battery will take to reach its end of life (EOL). In essence, SOH is a measure of internal resistance, capacity, voltage, self-discharge, the battery's ability to accept charge, and the total number of charge–discharge cycles that the battery has completed at that point in time. The state of health calculation is an algorithm that is programmed into the battery management system main controller.

State of Life (SOL)—The SOL is often used synonymously with the state of health. However, it may also be a separate algorithm that is intended to determine how much life is remaining in the battery based on some of the same measurements as are used in the state of health calculation.

SOx—The term SOx is used to represent all of the "State of" calculations at the same time, including state of health, state of charge, state of life, etc.

Voltage (V)—Voltage is the potential of the charge in a battery, but for clarity it can be thought of as being analogous to the pressure in a hose.

Now we have a base understanding of the terminology and lingo that are used in battery design. Chapter 4 will begin discussing some of the basic calculations, questions, and methodology for doing a high-level battery concept development.

Battery Pack Design Criteria and Selection

Perhaps one of the most challenging and frightening aspects of battery design, especially for the non-engineer, is trying to make the initial calculations around a new battery solution. However, there are a few relatively simple calculations that can be made that will allow you to appropriately size a battery concept for your application.

The first thing that must be understood is that a lithium-ion battery pack is a system of interrelated subsystems that are all necessary to make the battery pack function and meet its life requirements. At the core of the pack are the lithium-ion cells. The amount of cells will vary from one application to another, but every battery pack will require a group of cells connected in different ways in order to achieve the voltage and energy desired. In order to "hold" and manage the cells a mechanical structure is put in place. This mechanical system will include both some way to connect and hold the cells as well as the overall enclosure (Figure 1).

Within this enclosure, you will also find the battery management system (BMS). The BMS is an electronic controller that monitors and manages all of the functions of the battery operation. The BMS may also have separate electronics installed at the cell or module level which monitor the temperature and voltage of the cells, often referred to as a voltage temperature

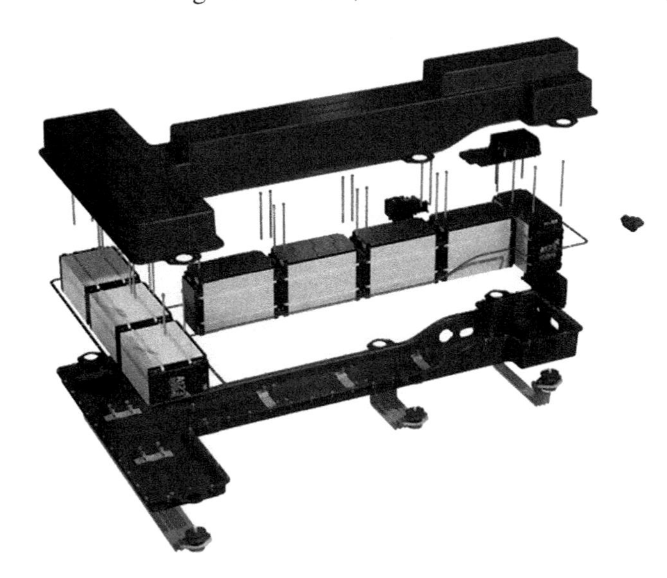

Figure 1
A123 lithium-ion battery exploded view.

The Handbook of Lithium-Ion Battery Pack Design. http://dx.doi.org/10.1016/B978-0-12-801456-1.00004-X

monitor (VTM) board. In addition to these there will be a thermal management system, which may range from a passive solution such as using the enclosure as a thermal heat sync to an actively managed liquid- or air-cooled system that forces cooled (or heated) air or liquid through the battery pack. And of course let us not forget the electronics, switches to turn the current flow on and off and wiring. All of these different systems must come together into a single system solution in order for the battery to be functional, safe, and meet its life and performance targets.

And while I will not spend much time trying to give cost or price estimates as there are many others out there who have continuous research on this aspect, I will offer up some thoughts as to what percentage of the battery these components make up. Note that for small power-based battery packs the cells make up much smaller percentage of the total pack cost, than compared to an electrical vehicle (EV) or a plug-in hybrid electrical vehicle (PHEV). This is also true of very large grid scale type systems, where the hardware around the cells may make up a larger percentage of the overall cost. In the PHEV/EV applications batteries can make up 60–70% of the complete battery pack costs (Figures 2 and 3).

Now that we have a basic understanding of the parts of the system, the next thing to understand when sizing a battery is what type of application it is? A battery is basically an electrochemical energy storage device. And we can divide battery usage into two categories: either power or energy. A power battery can be described as one that is focused on providing "instant" power for vehicle acceleration. The power battery is not intended to provide long-term energy to operate a system. Think of a power battery as one that discharges its power and energy in very small time frames, usually from seconds to tens of minutes. The energy battery is intended to provide a long and relatively slow discharge. Think of it as one designed to discharge its energy over 1 h or longer. In an automotive application this would typically be

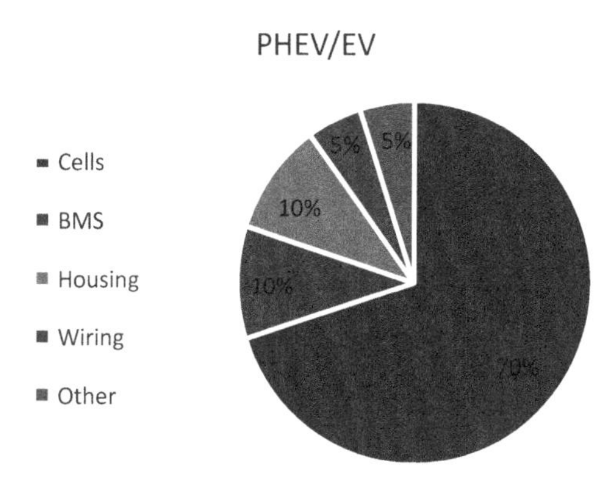

Figure 2
PHEV/EV battery cost breakdown. BMS, battery management system.

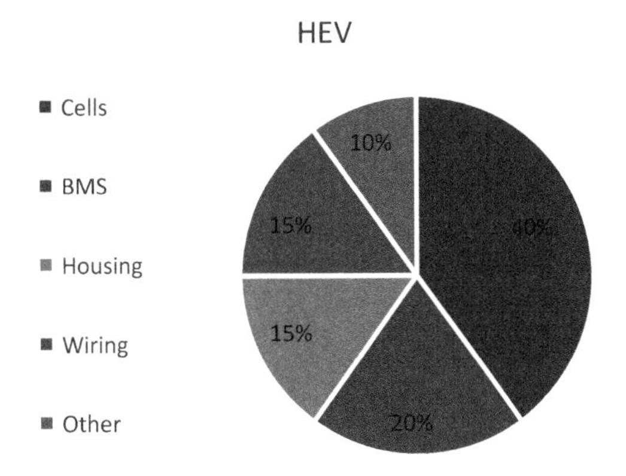

Figure 3
HEV battery cost breakdown. HEV, hybrid electrical vehicle.

found in a fully electric vehicle that is designed to achieve long driving ranges. In a grid type application this may come through as a battery backup system designed to provide power for several hours. There is also a third type of application that is not frequently talked about. The typical PHEV uses what I refer to as a "balanced" battery. In this type of application the battery needs to operate some of the time like a fully electric vehicle and at other times like a traditional hybrid. In this case the battery needs to meet both performance requirements.

A power battery will have a relatively small amount of energy, usually less than 1.5 kWh. An energy battery needs much higher energy and will range anywhere from 7.5 kWh up to 80 kWh or even greater for very large applications, with the average automotive EV battery being about 24 kWh.

In Appendices A through E I have included the battery goals that have been set out by the US Advanced Battery Consortium. These targets are set out for 12 V stop/start type batteries, 48 V stop/start micro-hybrids, and higher voltage HEV, PHEV, and EV type battery systems.

In all applications batteries must be connected with a combination of cells in series, to achieve system voltages, and in parallel, to achieve capacity. The trick is figuring out first what voltages the application needs so that you can back calculate to the required amount of cells to achieve that voltage; of course that assumes that you already know what type of cell you are going to use or at least the general chemistry. Finding the energy requirements can be a bit more difficult. In transportation applications if the efficiency of the vehicle is known it is easy enough to back calculate the desired energy, but for other applications it isn't quite as easy as that.

Another important thing to understand about a lithium-ion battery is that it will naturally lose capacity over its life due to high temperatures, an increase in internal impedance, and other factors that will determine how long the battery will actually operate. And while lithium-ion does not have a "memory effect" as some of the nickel-based chemistries do, the continued impedance growth will eventually reduce the capacity to a level where it is no longer fit for use in the applications.

Lithium-ion batteries will also "self-discharge" or loose energy as it sits in storage. There are two basic forms of self-discharge, permanent and temporary. Permanent self-discharge means the battery will never be able to return to its original capacity. This is usually due to an increase in the impedance within the cell while it sits in storage. Temporary capacity loss is that capacity that is lost during the storage period but will return once the battery is cycled again. Different chemistries have different performances in this category.

In the next sections we will begin delving into the basic calculations that are used in battery design. Most of these are derivations of Ohm's Law. Based on experimentation done by Georg Ohm back in 1825 and 1826, Ohms' Law basically states that current is approximately proportional to electric field, in other words voltage is equal to the current multiplied by the resistance of a circuit (Ashby, 2009). This fundamental law forms the basis for almost all of the basic calculations that follow. This section will use this basis to define some of the core calculations described in simple terms that can be used in order to determine things like how many cells are needed in a battery, what capacity cells are needed, and so on.

Ohm's Law and Basic Battery Calculations

While there are many different formulas that can help you in your battery pack sizing calculations, perhaps the most important is Ohm's Law. Ohm's Law describes the relationship between voltage (V), current (I), and resistance (R) in an electrical system (Ashby, 2009). And since voltage and current are two of the few things that we can actually measure in a battery (other than temperature, almost all other factors are calculated) they become critical in our calculations. Ohm's Law then is stated as such:

Voltage (V) is equal to current (I) multiplied by resistance (R)

Which can be described as:

$$V = I \times R$$

Now the beauty of this equation is that from this one equation, if you have two of the three variables you can calculate the missing one. For example, the same equation rewritten to calculate current as current (I) is equal to voltage (V) divided by resistance (R):

$$I = V/R$$

And the third part, as you may now have guessed is that you can calculate resistance using the same functions. Resistance (R) is equal to voltage (V) divided by current (I):

$$R = V/I$$

In the previous Chapter, I defined these terms but let me now give you a visual example of the three in order to help you better understand them and know how they interact. The graphic in

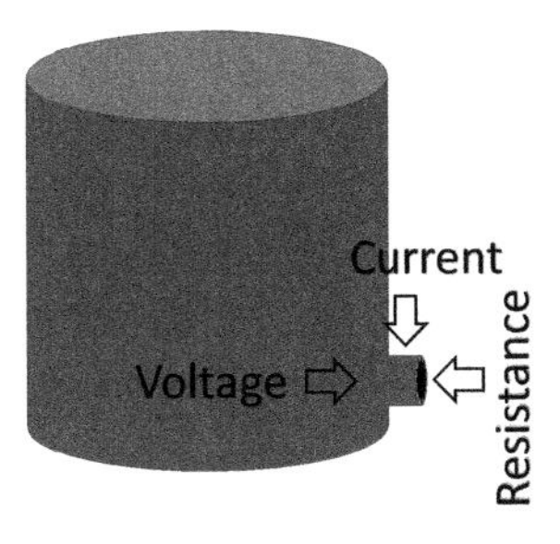

Figure 4
Ohm's Law described.

Figure 4 above is simplistic, but should do to help us understand these three items and how they are interrelated. In Figure 4 we see a large water tank, the size of the tank is analogous to the amount of energy that is in the battery system. The voltage is analogous to the amount of water pressure at the bottom of the tank that is forcing the water, or in this case, the electrons forward. The current in this example relates the size of the pipe through which the water is flowing out. So the larger the current the more power can be released. And finally the resistance is a bit more difficult to describe, but let us think of it as the friction that exists within the pipes, which tends to cause the current to run slower

This can also be described in electrical engineering terms as an equivalent circuit model, which is shown in Figure 5.

As you continue on in this chapter you will quickly see that this set of formulas becomes critical to understand in order to calculate virtually all of the other formulas that follow.

Note that in the following calculations I am generally using nominal values at the beginning of life (BOL) unless otherwise stated. In most cases this is adequate for the basic battery pack sizing exercise. Once you begin delving into the details of the operation of the pack, you can begin applying factors such as internal resistance, depth of discharge, state of charge, temperature, and other factors to accurately size and determine the end of life (EOL) of a battery system; but as our purpose here is to provide only a brief and simple overview we will focus only on the "basics".

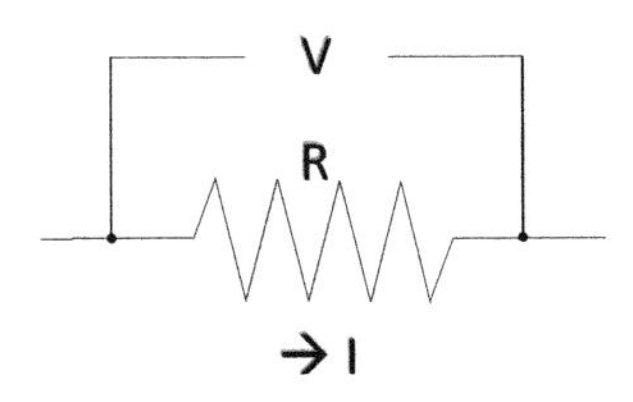

Figure 5
Equivalent circuit model.

Calculating the Number of Cells Needed

The first thing we need to do is to figure out how many cells will be needed to meet the basic voltage and current requirements of the pack design. Beginning with the desired pack voltage you can easily calculate the number of cells (assuming that you have a desired cell to work with) needed to create that system voltage. The system voltage is generally based on the voltage needs of the electric motor(s) in the system. Once this target voltage is known it is a simple enough matter to determine how many cells are needed in series to generate that voltage by dividing the pack voltage (V_p) by the cell voltage (V_c):

$$V_p/V_c = \text{number of series cells}$$

For example, for a desired 350-V pack and using a 3.6-V NMC (lithium nickel manganese cobalt oxide)-based cell, it would require 350 V/3.6 V = 97.2 cells, for simplicity round down to 96 cells. Similarly, to achieve a 350-V pack using a 3.2-V LFP (lithium iron phosphate) cell you would need (350 V/3.2 V) 109 cells. Again for simplicity you may decide to use either 108 or 110 cells. Or using a 2.2-V LTO (Lithium Titanate) cell you would need (350 V/2.2 V) 159 cells, or rounded to 160 cells in order to reach the desired voltage.

One of the reasons to round up or down here is also to end up with an even number of cells which will allow you to divide them equally in the modules that will make up the pack. This also allows you to use a single, or at least a common, module design in your pack. Going back to the NMC example, with 96 cells required we could manage this with 8 modules of 12 cells each or 4 modules of 24 cells each. If we rounded up to 98 we would have to use 7 modules of 14 cells each.

The other consideration that must be taken into account when determining how many cells and modules will be used in the design is the type of cell supervision circuit (CSC), also called a voltage temperature monitor (VTM), board that is used and how many cells it can monitor. The CSC will be discussed in more detail in Chapter 8 on Battery Management Systems, but for the purposes of this chapter we can say that it is important to know how many cells your CSC boards can manage; current technology ranges from 12 to 16 cells. So if your CSC can only manage 12 cells, then the module must include no more than 12 cells which means we will need 8 modules to achieve the 96 cell total and 350-V pack voltage level.

Calculating the Pack Energy and Capacity

To then calculate the pack energy (E_p), let us assume that a 25-kWh pack is desired, we would multiply the pack voltage (V_p) by the capacity (I_p). This of course assumes that you know the needed capacity, or current, of the pack.

$$E_p = V_p \times I_p$$

In this example, let us assume that we are using a 3.7-V nominal NMC cell and 96 of them in series to get a pack Voltage of 355 V and in order to achieve the desired energy we are using 70 Ah of cells. This could be accomplished by using two 35 Ah cells in parallel or one 70 Ah cell, or some other parallel combination that leads up to 70 Ah. So we get:

$$355.2\ \text{V} \times 70\ \text{Ah} = 24.864\ \text{kWh}$$

Also keep in mind that the required capacity is generally determined by the electric motor current requirements and energy is typically determined by the end user based on the application type. For instance, if the end-use application is an all-electric vehicle that is targeting about 75 miles of driving range with a system that is capable of providing 3 kWh/mile efficiency, then:

$$75\ \text{Miles} \div 3\ \text{kWh/mile} = 25\ \text{kWh battery size}$$

Also note the units used in these calculations, I have taken the liberty of using 25 kWh in my calculations instead of 25,000 Wh. But if you are running these numbers on your calculator along with the book you will find that the results do not match and the decimal point is off! To be fully accurate in your calculations you will find that you need to use 25,000 Wh/350 V to get the 71 Ah target. For the sake of brevity I am making that calculation automatically for you here.

Now one last thing to note here goes back to my earlier comment about how much current the electric motor or system will need to draw. A 71 Ah cell will by definition be able to supply 71 A for a 1-h period and will be able to offer higher power pulses at different rates. But this may not meet the current requirements of the overall system. Many PHEV and EV systems for instance may require sustained current draw of 30 A, 40 A or up to around 400 A depending on the application and the system design. So in order to achieve, for this example, a 90 A continuous current (89 A rounded up) you would need to either use a single 90 Ah cell, two 45 Ah cells in parallel or some other combination of lower Ah cells in a parallel configuration in order to achieve these targets.

This leads us to a brief discussion on "C-rates." A C-rate is a measurement that is provided by the cell manufacturer to describe how much current the cell is capable of providing over some period of time (t). So if you have a cell that is 70 Ah at a 1C rate it will provide 70 A of capacity over a 1-h period. But that cell may be rated up to 5C for a period of 10 s which means that it would provide 350 A (70 A × 5) over a 10-s period. The same calculations can be used for charging and regenerative braking. The battery must be able to accept the C-rate charge that the system will deliver otherwise the energy will not be accepted in the battery and is often transformed into waste heat during regenerative braking for instance.

Alternatively, if you know how many cells are to be used you can multiply the total number of cells times the energy in watt hours of each cell to get an estimate of the total pack energy. This can be a useful tool if you conduct a packaging exercise and are able to determine how much room for cells you have in the system. It is not uncommon to do a quick packaging study to find out how much room is available for the battery. With this information you can quickly estimate how many cells will fit into the space and determine what size battery you can offer.

For example, if we are using 96 Cells of 259 Wh per cell (3.7V × 70 Ah per cell) it equals 24,864 Wh or 24.9 kWh of total pack energy. Again, this also assumes that you have a specific cell in mind to be used in this application. This method is best if you already have a portfolio of cells that you can use rather than if you are trying to determine which cell is best. The prior method is better suited to determining what size cell is needed.

Now I would like to just quickly relate what we have reviewed already back to Ohm's Law here. We looked at a couple of formulas for calculating pack energy and voltage using Ohm's law. Remember that Ohm's Law states that R = V/I. In this case we have used this same formula to calculate the following three formulas for capacity (current), energy and voltage:

$$24.864 \text{ kWh}/355.2 \text{ V} = 70 \text{ Ah}$$

$$355.2 \text{ V} \times 70 \text{ Ah} = 24.864 \text{ kWh}$$

$$24.864 \text{ kWh}/70 \text{ Ah} = 355.2 \text{ V}$$

So to quickly summarize what we have determined with this example, we need:

- 96 cells at no less than 3.7V in series in order to achieve the pack voltage
- One 70 Ah cell in order to achieve the pack current requirement of 70 A
- A total of 96 of these 3.7V and 70 Ah cells
  - Alternatively you could achieve the same system performance results using two 35 Ah cells in a parallel configuration and a total of 96 of these 2P cell groupings which would require a total of 192 cells to the system

Calculating Pack Energy at End of Life

Of course all of these calculations assume that you can use 100% of the battery to achieve that range, in reality you may only be able to use about 80–90% of that battery depending on the cell selection and usage profile. That means that the 25 kWh must be the amount of energy that is *usable* in this system design. In other words, we must be able to remove 25 kWh of energy from the pack and, therefore, if we are to maintain the safety margins on the top and bottom and only use 80% of the total energy, we will need to determine how much total energy is needed. That means another quick calculation by dividing the usable required energy by 80% (assuming here that the system will use only 80% of the total energy) gives us:

$$25 \text{ kWh}/80\% = 31.25 \text{ kWh } \textit{total} \text{ battery pack energy needed}$$

This means that in order to remove 25 kWh of energy from the pack to achieve the 75 miles range, the battery will really need to be sized up to over 31 kWh in order to achieve these goals. Of course the most important question to ask your customer when talking about pack energy is which number they are looking for usable energy or total energy. If the customer has already calculated the efficiency of their vehicle then they may be referring to a need for a total energy of 25 kWh less the 20% that is not usable from the system design and not in fact 25 kWh of usable energy, which will require a larger system in order to deliver that much usable energy.

So with all of these things in mind, and assuming that we need the larger 31.25 kWh pack, we can recalculate the capacity of the cells in order to achieve these goals using the formula:

$$E_p/V_p = I_p$$

Where E_p equals the total pack energy in kWh and I_p represents the pack current in Ah, and V_p represents the pack voltage that we are calculating here. So using the formula for this example we find:

$$31.25\ \text{kWh}/350\ \text{V} = 90\ \text{Ah cell}$$

In this example, we would need 96 cells that are 3.7 V each and at 90 Ah in capacity to achieve the 350 V and 31.25 kWh of total energy. This distinction between total energy and usable energy is an important one because virtually all lithium-ion batteries available today are not capable of using 100% of their available energy in consideration with safety, life and performance requirements.

Calculating System Power

With those basic calculations in mind we can also dig a bit deeper and begin looking at how much power the system can provide. In addition to the formula's shown above which are based on Ohm's Law it is also possible to use several derivations of this to calculate power and to use power (in Watts) in your calculations. In this instance, we are combining Ohm's Law with Joule's Law in order to create a formula for determining electrical power (in Watts again). System power can be calculated by using the formula:

$$P = I^2 \times R$$

Where P represents the power in watts (W), I^2 represents the square of the current in Amperes, and R represents the resistance in Ohms. Similar to Ohm's Law this formula can be cross-calculated to form the following three base formulas for calculating power:

$$P = I^2 \times R$$

$$R = P/I^2$$

$$I^2 = P/R$$

For example, if the cell that we are using has an internal resistance of 7 mΩ, then we can calculate the power for each cell as follows:

$$(90\ \text{Ah})^2 \times 7\ \text{m}\Omega = 648\ \text{W per cell}$$

or

$$648\ \mathrm{W}/(90\ \mathrm{Ah})^2 = 7\ \mathrm{m\Omega}\ \text{per cell}$$

or

$$648\ \mathrm{W}/7\ \mathrm{m\Omega} = 92\ \mathrm{Ah}\ \text{per cell}$$

Using this first formula ($P = I^2 \times R$) we can also replace current with voltage to develop another set of formulas for calculating electrical power (again in watts):

$$P = I^2R = V^2/R = V\ (V/R)$$

Finally, these formulas can be reversed to calculate current (I) and voltage (V) in the same manner:

$$I = P/V$$

$$V = \sqrt{(P * R)}$$

In any of these cases you need to have the measurements for voltage, current, resistance, or power or a combination of them in order to make these calculations. An alternative method that may be a bit simpler to use, but not necessarily more accurate, is to quickly estimate the power in kW of a pack by multiplying the total number of cells times the power in watts (W) of each cell, for example:

$$96\ \text{cells} \times 648\ \mathrm{W} = 62,208\ \mathrm{W}\ \text{or}\ 62.2\ \mathrm{kW}$$

In this case the cell power is usually available from the cell manufacturer on their published data sheet. If not available on the data sheet, the cell manufacturer should be able to supply this information to you.

In addition to calculating the nominal power capabilities of your battery system, you should also be able to calculate the "peak power" that the system can provide. This is usually a 10-s, 5-s, 2-s or 1-s calculation. But the general formula is a two-part calculation.

First the system resistance must be calculated. This is typically done by running a hybrid power pulse characterization (HPPC) test on the cell and measuring the change in voltage and in current and then dividing the two:

$$\text{Resistance}_{DC} = \Delta V / \Delta I$$

Once you have the resistance you can use it in the following formula to calculate the peak power potential of the system:

$$\text{Peak Power (in kW)} = V^2_{\text{Open Circuit}}/4R$$

In this formula you are dividing the square of the maximum open circuit voltage by four times the resistance to find the peak power.

Maximum Continuous Discharge

In order to calculate the maximum continuous discharge current that the system can provide, you need to multiply the number of cells in parallel (N_p) times the cell current (I_c) multiplied by the maximum C-rate (C_{Max}) to get the maximum discharge current ($I_{Max\ Continuous}$) in amperes. The other method for this is to use the maximum discharge current that is supplied by the cell manufacturer on their data sheet multiplied by the number of cells in parallel to get a close estimation of the maximum discharge current.

$$N_p \times I_c \times C_{Max} = I_{Max\ Continuous}$$

or

$$1\ (\text{cell in parallel}) \times 90\text{A} \times 5\text{C} = 450\text{AMax Continuous Discharge}$$

This same formula can be used to determine the maximum continuous charge current that is available; simply replace the C-rate or the current with the maximum continuous charge current or maximum continuous charge C-rate that is supplied by the manufacturer.

Calculating Charge Voltages

Maximum charging voltage is simply equal to the total number of cells in series multiplied by the cell's maximum voltage as defined by the cell manufacturer.

$$96\ \text{cell} \times 4.2\text{Vmax} = 403\ \text{V Maximum}$$

Minimum is calculated in the same manner: number of cells in series multiplied by the cell's minimum voltage as defined by the cell manufacturer.

$$96\ \text{cells} \times 2.7\text{V} = 259\ \text{V Minimum}$$

Converting Customer Requirements into Pack Designs

Now that we have a basic understanding of the types of formulas that are needed in putting together a basic lithium-ion pack concept, we should talk briefly about customer requirement documents. Customer requirement documents can range from hundreds of pages for the more sophisticated customers to extremely simple one line descriptions for companies that are less familiar with battery technologies. But in either case the most important thing that should be done prior to beginning to develop your battery design and concept is to sit with the customer to discuss what the requirements mean and what the most important characteristics of the requirements are.

When you meet with the customer you will get a better understanding of what they are trying to achieve with the battery, which will drive your specification development. You may find that the inverter and the motor have different voltage ranges than were indicated in the requirement document. You may also find that you can optimize a system to meet a customer's specific key requirement. For instance, I have frequently found that it is possible to reduce the overall capacity of a battery pack by using a higher power cell for an application that really needs a power battery. In this type of case you may be able to offer a lower cost system to the customer that still meets their key requirements.

Another key part of the requirements that you may not actually get in the requirement document but will be able to get while speaking to the customer is a performance profile. In this case what you are looking for is the charging and discharging power and energy needs over a specified time period. Figure 6 below offers an example of this type of power profile; in this example we are looking at the power in Watts that is needed for an application over a 24-h time period. In this example the requirement may indicate that we need nearly 350 W of peak power, but in reviewing the operating profile we see that the application really only needs this much power for one short period each day and the average power demand is only 175 W. That means that we may be able to offer them a slightly smaller capacity battery, but using a higher power cell we can still meet these peak power needs.

At the other end of the requirements spectrum there are some requirements which may only tell you how much power or energy they need and not much more. For example I have seen some stationary and grid type battery requirements that only state that the system will need to deliver 15 MW of power for 15 min, for example. In many of these types of requirement documents it is impossible to fully develop a battery proposal based solely on this limited

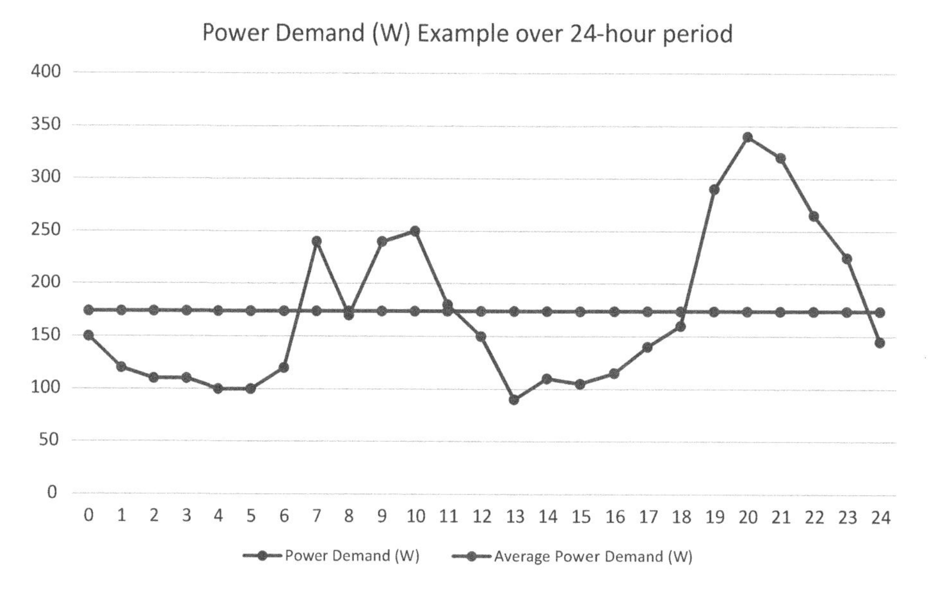

Figure 6
Sample power demand profile.

requirement. There are some relatively simple things you can assume here, for instance to deliver 15 MW of power for 15 min likely means that you need a battery that is about 60 MWh of capacity in order to deliver that much power for that period. This of course assumes a 1C discharge rate. However, we also need to understand how many charge–discharge cycles of this type are expected and how often. Perhaps the system only needs to deliver this much power a couple of times per year, or perhaps it needs to deliver this much power on a daily basis. In each of these two cases we could develop two very different energy storage systems to meet each of these two different cycle requirements. This is a good example of when it is important to conduct a technical review with the customer. For the example of a grid scale application you should also meet with the power electronics company that is providing the inverter and power electronics components for this system (assuming that you as a battery provider are not already providing these components). Meeting with this partner may give you much of the key information that you will need to properly size your battery. From them you will be able to understand the system voltage requirements, interconnect voltages, communications requirements, and many other important characteristics that you battery will need to meet in order to achieve the customer demands on the system. You may also need to spend some time with the engineering services firm that is doing the overall site design and layout to understand where the battery will be installed, will it be in an air conditioned building or will you need to design a trailer or container in order to house the battery system. If this is the case, you will also need to understand some of the local building ordinances and requirements and the geographic location for the installation. These may drive safety and fire detection and prevention systems to be installed in the container.

Power to Energy Ratios

Another topic that we should make sure to discuss is power to energy ratios. The power to energy ratio is a quick number that many customers and system designers use to quickly evaluate the fit of a certain technology for their application. High power applications, such as a 12 V stop/start type automotive battery, will typically have much higher power in relation to the amount of energy that is stored in the battery pack. In this case we may see a ratio between 15 to 1 (15:1) and 20 to 1 (20:1). This ratio represents the amount of power that the battery is capable of providing compared to the amount of energy that is stored on board.

If we continue using this example and base it on the USABC guidelines (see Appendix A) for our approximations, we find that the "ideal" 12 V stop/start battery should be able to provide up to 6 kW of power for 1 s, and it should also provide 360 Wh of energy. So in this case, 6,000 W/360 Wh (converting to similar units) equals a 16.7:1 power to energy ratio.

Table 1 below shows some average power to energy ratio ranges for different automotive applications; however, note that these same estimations can be made for all other energy storage applications.

Table 1: Power to energy ratios

12 V S/S	48 V	HEV	PHEV-20	PHEV-40	EV-100
15:1 to 20:1	25:1 to 40:1	30:1 to 35:1	6:1 to 7:1	3:1 to 4:1	2:1

From this table it is clear that the hybrid type applications have a much higher power need than the energy applications do. This also validates the comment that is frequently heard that with a large energy storage systems that power is much less of a concern because it is already there.

Of course, that does not mean that we should not take care in our cell selection when sizing a large battery. As has been stated previously we may find that we can reduce the overall capacity of the battery if we can get a high enough power cell thereby reducing the customer cost and optimizing the system to meet the customers actual performance needs.

Large Stationary and Grid Systems

These same calculations, formulas, and processes that I have included above can be used in evaluating and sizing a battery-based energy storage system for a large grid or stationary system. The challenge for most battery manufacturers is that the amount and level of detail of the requirements from these types of applications are typically much less than would be expected from one of the large automakers.

Examples of this may be seen in some of the recent request for proposals (RFPs) for energy storage systems for grid and renewable integration applications. The author has seen many RFPs which require energy storage of "60 MW for 30 min" as the entire battery specification. From this the only real sizing that we can do is to convert it to energy: If they need 60 MW of power for 30 min, then we can convert it to energy that would be 120 MWh (60 MW × 2). More than that we need to refer back to the section on converting customer requirements and we need to meet with the power electronics equipment suppliers to determine what voltages their inverters and systems need to run at and what other requirements they may have. In general, the less detailed the requirement documents are, the more work and time is needed to be spent with the customer in order to gain a full and deep understanding of what they are trying to achieve with the system.

Quick Formula Summary

Just like in your basic algebra class in high school, as long as you have two variables you can calculate the third. Below is a list of the calculations that were included in this chapter. With these formulas you should be able to make most of the basic calculations necessary to get a beginning understanding of your energy storage system's performance.

Calculating voltage (V):

$$V = I \times R$$

$$V = \sqrt{(P \times R)}$$

Calculating current (I):

$$I = V/R$$

$$I^2 = P/R$$

$$I = P/V$$

$$I_p = E_p/V_p$$

Calculating resistance (R):

$$R = V/I$$

$$R = P/I^2$$

Calculating power (P):

$$P = I^2 \times R$$

$$P = V^2/R$$

$$P = V\ (V/R)$$

Calculating Energy (E):

$$E = V \times I$$

Number of series cells:

$$\text{Series cell} = V_p/V_c$$

Design Guidelines and Best Practices

Perhaps the most important aspect of beginning to develop a new battery pack design is ensuring that you have a complete view and understanding of the customer's actual requirements. Many of the largest automotive original equipment manufacturers will generate very detailed and complex requirement documents. This is the best scenario as it will give you a complete understanding of all of the various requirements.

On the other hand, you may find yourself working with a small customer that does not have a set of requirements and you will be forced to work closely with them in order to gain a full understanding of what they hope to achieve with the battery. At this stage it is important to ask as many questions as possible. I often find it valuable to begin by asking about the electric motor power requirements (how much power is needed) followed by the electric run time requirements (how long is it expected to run). Once you begin collecting this type of information you can begin using some of the calculations above to help size the battery to meet the customer expectations.

CHAPTER 5

Design for Reliability/Design for Service

One of the great challenges in designing a large lithium-ion battery is estimating and calculating the reliability and lifetime of the energy storage system. This is in large part due to the fact that there is not yet enough history on this technology that is available to be able to base future predictions on past performance. Therefore engineering fields like design for reliability (DFR) become very important very early in the engineering process in order to attempt to identify potential failure modes and develop mitigation strategies.

With automotive applications like the Nissan Leaf and Chevrolet Volt being on the market for about five years now, the batteries in these products are very likely only about halfway to the end of their useful life in the vehicles. The best estimates that we can draw from today are based on the many nickel metal hydride (NiMH) hybrid electric vehicles that are on the road today. Having been introduced in the late 1990s many of these are now beginning to reach the end of the vehicle's life, typically about 12 years in the United States. So while there are some learning that can be made from these vehicles, most of the automakers and battery manufacturers are being very conservative in their designs so as not to incur large warranty costs if these energy storage systems reach the end of their useful life before they reach the end of their warranty period. They are using methodologies such as DFR in order to help build their confidence in the longevity of the energy storage systems.

There are several relatively recent examples that exemplify the need for battery manufacturers to implement both reliability and quality systems. The first and perhaps the biggest recall that happened in the lithium-ion battery area was conducted by Sony in 2006 when over 10,000,000 battery packs were recalled at an estimated cost to Sony of about $429 million USD (Arendt, 2006). The reason for this particular recall may be attributed to a combination of both product and process failure modes. In this case, the 18650 type lithium-ion cells (18 mm diameter and 65 mm length) that were recalled used a nickel coated steel can with a lid that was crimped on. Subsequent analysis indicated that during the crimping process small particles from the nickel may have been flaking off and falling into the jelly roll assembly, which in turn had the potential to cause an internal short circuit in the cell over time. In essence this was a latent failure that showed up only over time.

Specific to the automotive industry, lithium-ion battery start-up A123 Systems ended up recalling more than $55 million in batteries that were introduced in the Fisker Karma (Voelcker, 2012). This was the second recall for A123 who suffered an earlier issue with coolant leakage in the Fisker battery pack. As an early start-up company these recalls were

The Handbook of Lithium-Ion Battery Pack Design. http://dx.doi.org/10.1016/B978-0-12-801456-1.00005-1

very damaging to the company and the industry as a whole, because it was one of the major drivers that led A123 Systems ultimately filing bankruptcy in the US. This second failure was a process failure, in that it is believed that there had been a miscalibration problem with one of their welding machines during the cell production process which created the possibility for latent failures in the cells in the future (Rousch, 2012).

Now these are both good companies with very good products, and there have been other expensive recalls in the battery world so do not let me lead you astray in thinking that these are not quality products, but these examples emphasize the needs for robust quality control and reliability planning in the battery industry, as the industry is in very early stages of commercialization.

Design for Reliability/Design for Service

Much like all other engineering fields, DFR and design for service (DFS) are two important processes that must be integrated very early into the lithium-ion electronic safety and security design process. DFR is the use of a systematic and concurrent engineering process which is integrated throughout the entire product development cycle and is focused on achieving product reliability and durability over the products lifetime. Described in the *Reliability Edge* quarterly newsletter, DFR is

> *...a systematic, streamlined, concurrent engineering program in which reliability engineering is weaved into the total development cycle.*
>
> ***(ReliaSoft, 2014, p. 1)***

DFR is not a single process but rather a series of processes such as failure modes effects analysis (FMEA), test to failure (TTF), accelerated life testing, voice of the customer (VOC), and design of experiments (DOE) among many others. The main challenge with implementing these tools is that they are mainly focused on using historical data in order to predict future results, and with lithium-ion battery systems that presents some challenges as there is very little historical data on systems that are this large. While there is a large amount of data on lithium-ion battery in portable power products like laptop computers and cell phones it does not always scale up directly to the larger systems.

DFS essentially involves evaluating the designs at the very early stages looking for solutions to improve the serviceability of the energy storage system. In some cases this may involve using a mechanical fastener to interconnect the cells. This allows for the ability to replace a single cell, thereby making it the smallest replaceable unit (SRU). However, there are also challenges with this strategy that must be evaluated, such as the risk of the fasteners loosening over time and creating a failure mode in that manner. Other battery manufacturers look at welding the cells together instead of using fasteners. This approach makes the "module" the SRU, making it a much more expensive part. However, by welding the cells, there is a smaller likelihood that the cell interconnections will fail over time. The other burgeoning industry that a DFS engineer should take into account is the recycling, second-life and remanufacturer. As this industry begins to take shape it will become even more important to ensure that your energy storage systems can be easily broken down and each of the core components easily identified.

So in essence, DFS involves looking at who will be doing work on the battery pack and trying to develop products that ensure the ease of service. And DFR involves attempting to identify all of the potential manners in which an energy storage system could fail in order to engineer in appropriate mitigation strategies.

Quality and Reliability

Perhaps at this point we need to make a couple of clarification points. Specifically, what is the difference between reliability and quality and how are they related? Most simply, quality control looks to ensure that the product will work as intended and meets the assembly and manufacturing specifications. So quality is concerned with *how* the product will work. Reliability on the other hand is concerned with *how long* the product will work as it was originally designed. In essence, reliability provides the statistical probability that the product will continue to meet its functional purpose throughout the design life (ReliaSoft, 2014). The following chart demonstrates the differences and the areas of overlap between quality and reliability, showing the different processes and tools used for the quality system from design for six sigma and the reliability programs from DFR (Figure 1).

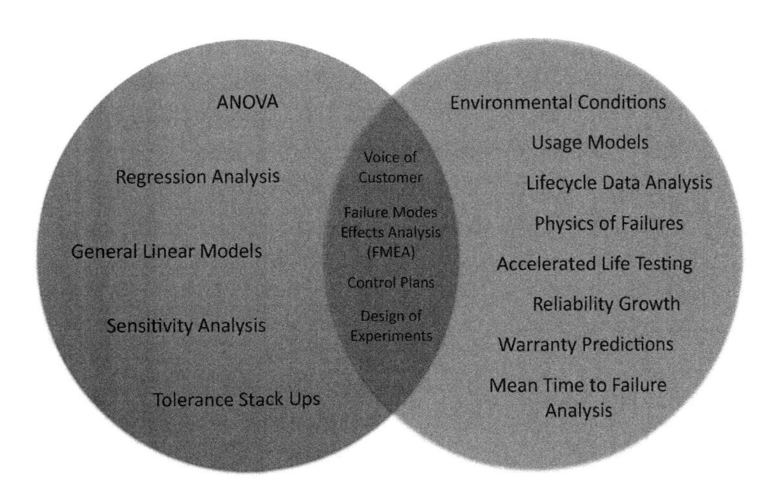

Figure 1
Overlap of Design for Reliability and Design for Six Sigma. *Adapted from ReliaSoft (2014, p. 3).*

Failure Modes Effects Analysis

One of the tools that is used for both quality and reliability analysis is the FMEA. The FMEA is a tool that is designed to systematically evaluate potential ways in which a product or process could fail, to assess the risk of those failures occurring and finally to attempt to prioritize those areas where corrective actions are needed in order to mitigate those failure modes as well as to document the corrective actions that are implemented as a result of the FMEA and their results (American Society for Quality, 2014; Automotive Industry Action Group, 2014). The FMEA is generally broken into a couple of different areas: concept FMEA (CFMEA), design FMEA (DFMEA), and process FMEA (PFMEA). The difference between them is only based on the area of focus and the timing during the development process. The CFMEA is focused on identifying the potential failure modes in the concept before the actual hardware is defined. More commonly done is a DFMEA which is focused on the product itself and identifying the potential areas where the product may fail. For example, a DFMEA for a battery system could include an item such as "Potential for fasteners to loosen during vibration" which may cause a potential failure of, depending on the location of the fastener, a reduction in power, a potential to create an external short circuit, and an increase in heat generation and impedance due to the loose connection (these all assume that the fastener is used to connect a bus bar to a cell in this instance). The DFMEA then will evaluate the questions of how likely is this to occur, how often is this expected to occur, and what are the results of this failure. It may then drive a discussion around the design which could result in the bus bars and cells being redesigned to use a welded terminal (thus eliminating the potential for a fastener to loosen up). This is just a simplified example to help illustrate the DFEMA and how it works.

On the contrary, the process and assembly FMEAs are totally focused on the process used to manufacture the product. While this PFMEA could drive requests to the product team to make design changes, it is really focused on ensuring that the process is in place to ensure that the product, an energy storage system in our case, is assembled correctly and done in such a way as to minimize the potential for errors to occur during assembly. In this instance, the example may include a potential failure mode of a fastener not being fully driven or seated. The effect of this may be similar to the DFMEA example above; if it is for an electrical connection, a loose fastener may reduce contact between the bus bar and the cell terminal, causing a "bad connection" and a potential for a shortened life. In this instance, the actions that are recommended are likely to be more process focused such as adding a torque detect feature during the installation to ensure that the fastener is fully seated. But it may also include the addition of features into the design that enable error proofing during the assembly process. Figure 2 shows the industry standard FMEA form. Organizations such as the Automotive Industry Action Group (AIAG) and the American Society for Quality (ASQ) offer standard FMEA tools and even offer training and certification on using the tool.

Process Step / Function — Requirements	Potential Failure Mode	Potential Effect(s) of Failure	Severity	Classification	Potential Cause(s) of Failure	Occurrence	Current Process Controls (Prevention)	Current Process Controls (Detection)	Detection	RPN	Recommended Action	Responsibility & Target Completion Date	Action Results				
													Actions Taken & Effective Date	Severity	Classification	Detection	RPN

Figure 2
Sample failure mode effects analysis.

The FMEA is most effective when it is based on a product that is in the market already. In this manner, you will already have a good idea of the types of failures that can or have occurred. However, with lithium-ion energy storage systems, for large applications there is little history from which to base your evaluations and so must be based on "expert" judgment of the engineering teams involved.

The final thing I will mention on FMEAs is that they are a process that takes some time, dedication and knowledge to undertake. A typical DMFEA for instance may take several weeks or more to complete and is likely to be done in parallel to the design and engineering processes. Often they are also led by a quality engineer who is a member of the design team who is an expert in conducting FMEAs and may be certified by one of the organizations that I mentioned earlier.

Design for Service

DFS is a parallel activity to DFR that is intended to integrate service features and lifetime serviceability into the product. The DFS process is focused on integrating features into the product that can be used by service and support personnel to repair a product without having to replace it entirely. In the design process, DFS involves evaluating what components may fail, or are designed to fail first, and making them easily and safely accessible for service personnel.

Part of the DFS process may entail conducting mean time to failure analysis on the components and systems of the energy storage system in order to understand how long they may last before catastrophic failure occurs. This is already a common process in printed circuit board design as well as in much of the electronics industries.

The question when it comes to battery service life is what components should be serviceable. Many lithium-ion cells are connected together using a welded bus bar-type design. This means that once these are welded together they become a single unit, the SRU mentioned earlier. So in most cases the battery cells are not serviceable. There are some companies that have developed mechanical connections for the battery cells, however, in automotive applications there has been concern about meeting the reliability life of the product with these designs. Yet for other markets that do not have such challenging shock and vibration requirements, mechanical solutions may be perfectly acceptable solutions.

Frequently in the design process the battery module, the assembly of lithium-ion cells connected mechanically and electrically, is much easier to replace than replacing single cells. However, the cost of replacing a module will always be higher than the cost of replacing a single cell. So there is always a trade-off that must be evaluated during the design process. Another challenge that must be considered in servicing of battery modules is the aging of the lithium-ion cells. As a battery ages during normal use the capacity begins to shrink due to natural factors mentioned in an earlier section. Therefore when a battery module is replaced, it should be done with a module of the same capacity and energy. That means that you cannot replace an old module with a brand new one as the new module will have higher capacity and energy than the aged one. The problem with doing this is that the new module will always be limited in its performance by the weakest cells in the system, so the new module will never be able to operate at its highest performance capability. Optimally, the module should be replaced with one of equal capacity which can be achieved either by sorting "returned" modules and remanufacturing them so that they can be used as replacements or by artificially aging new cells to the same level as the ones in the module being replaced.

Other system components that must be designed for serviceability include things like fuses, controllers, electronics, and fans. Many of these components may be mounted in a serviceable compartment that allows them to be accessed without having to open up the complete battery pack and exposing the high voltage system. Fuses may be integrated into a manual service disconnect for larger PHEV or BEV battery packs but may be mounted into a separate fuse holder for small packs. In both cases they should be able to be accessed from outside of the pack. Controllers are also important to locate in places that allow them to be accessed in the event that a controller fails. For air cooled systems fans also present certain problems as most fans have lives of about three to five years. With the longer battery pack design and warranty life, fan selection can drive costs significantly for smaller systems. In order to achieve a 10-year design life in an automotive application it may be necessary to select more expensive fans or trade-off with a shorter fan design life.

Chapter Summary

Both DFR and DFS pose challenges to the lithium-ion battery design due to the fact that lithium-ion batteries for automotive applications have not been in mass product for very long. The first lithium-ion batteries were commercially produced in 1991 for the portable power market, but it was not until about 2009 that the first lithium-ion batteries began hitting the market for automotive applications. There simply are not enough data on lithium-ion battery performance in the automotive, stationary, industrial, or commercial fields to fully understand all of the facets of battery life. In addition to the lack of time in the market, many of the large applications are using "large format" lithium-ion cells that range from 10 Ah in capacity up to several hundred Ah in capacity which have only been introduced to the market in the around 2009 or so.

However, through advances in cell technology, engineering, and testing, a battery can be designed to last the life of the application. The design life for an automotive quality lithium-ion battery ranges from 10 to 15 years. This differs from the warranty period which is generally 6–8 years for most applications. Since lithium-ion batteries are being designed to last the life of the vehicle, currently about 11–12 years in the United States, the battery design must be robust enough to achieve these life targets. A design life of 15 years may be required to meet the California ZEV requirement, which requires a 10-year warranty and 15-year design life for all emissions-related components. Industrial and grid based applications may require much longer life periods from 15 to 20 years or more in some cases.

Design Guidelines and Best Practices

- Design for Reliability and Design for Service are engineering processes that are critical for ensuring the longevity and life of the energy storage system.
- Organizations such as the Automotive Industry Action Group (AIAG) and the American Society for Quality (ASQ) offer both training and templates that can assist an organization in implementing quality programs.
- The FMEA is an important process that should be conducted in parallel with the core engineering and design processes and should be conducted over a period of time.
- Lack of history and energy storage systems in the field demand the expertise of the engineering team in evaluating potential failure modes.

CHAPTER 6

Computer-Aided Design and Analysis

One aspect of battery design that is evolving quickly is computer-aided design, computer-aided engineering, and advanced analysis focused on battery systems. These tools can take many forms, from 3D models of the battery packs, which are common in the engineering field today, to computer simulations of the amount and location of heat that will be generated under a specified load profile, and simulations on the mechanical performance under different vibration profiles. There is also significant work going on to develop simulations of the different electrochemistries, the physics that occur within those chemistries, and how they will function and interact under different profiles and conditions. All of these tools are intended to help improve the battery design function and speed the development process time.

When we talk about sizing batteries, there is not any industry standard "calculator" that will be able to help, at least not for lithium-ion batteries. Several of the battery cell manufacturers have developed their own tools, some of which are available on the Web but often only for customers and they will only include their own cells in the comparison. Many of these companies develop these as tools for their sales and applications engineering teams to use in order to be able to quickly determine the size and capacity of a potential battery system using the basic theories of Ohm's Law that were described in Chapter 5. These are relatively easy to create in MS Excel or a similar spreadsheet or database type program and if you have enough performance data you can end up with a tool that is quite accurate.

Organizations and Analysis Products

When it comes to analysis tools, there are several available including some ongoing development tools that are being spearheaded by organizations such as the National Renewable Energy Laboratory (NREL) in conjunction with several different companies and partially funded by the U.S. Department of Energy (DOE). One such tool is the computer-aided engineering for electric-drive vehicle batteries (CAEBAT) project, which aims to accelerate the development and reduce the cost of lithium-ion batteries for next-generation electric-drive vehicles by achieving four main targets, which are as follows:

- Developing engineering tools to improve the design of lithium-ion cells and battery packs
- Reducing the timing required for battery prototyping and manufacturing
- Improving the overall battery performance, safety, and life span
- Reducing battery costs (National Renewable Energy Laboratory, 2013)

The Handbook of Lithium-Ion Battery Pack Design. http://dx.doi.org/10.1016/B978-0-12-801456-1.00006-3

Private companies, such as software analysis developer CD-ADAPCO who is also one of the participants in the CAEBAT program, already offer several analysis tools that are designed to help in the evaluation of different cell types, chemistries, and thermal performance but their solutions are currently targeted at the cell-level development.

In addition to the private companies developing software and analysis tools, many of the government labs and universities have been deeply engaged in developing tools. One such group is the Impact and Crashworthiness Laboratory at MIT. The MIT lab has been working to develop models in order to evaluate the effects of vehicle crash situations on the components within the cells, the cells themselves, and the battery pack overall. Their work has focused on understanding the effects of crash and penetration on each of the different types of battery cells through modeling, simulation, and then correlating that to actual testing. In a recent research paper published in *The Journal of Power Sources* (Zia, Wierzbicki, Sahraei, & Zhang, 2014), the MIT team presented their findings and analysis of the influence of road debris impacting battery packs mounted underneath the vehicle. This research was done in response to several incidents involving Tesla electric vehicles wherein the vehicle ran over debris in the road which subsequently penetrated the battery and caused failure of the pack. Through the use of several different modeling and simulations, the MIT team was able to develop a set of recommendations to assist vehicle and battery designers in first understanding the results of these types of events and then ensuring the safety of the battery when these events occur (Xia, Wierzbicki, Sahraei, & Zhang, 2014).

Another provider of software tools that is capable of this type of analysis is ANSYS. ANSYS offers a suite of software analysis tools called "Fluent," which is an advanced computational fluid dynamics (CFD) tool. While the tool was originally designed for analyzing mechanical and thermal systems, it does a very good job of analyzing the thermal management systems in energy storage systems. In this software package, the battery is modeled and the material properties and the heat generation of the cells are identified and associated to their respective components. The software then models the flow of either cooling (or heating) air or liquid within the battery pack in order to evaluate its effectiveness at cooling the battery as well as to analyze "hot spots" (ANSYS, 2014). This is a very valuable tool that can be used in designing the thermal system by modeling, modifying the design, and then running the model again and repeating this process in order to develop an optimized thermal solution. Another tool that may be evaluated to use for lithium-ion battery cell and system thermal analysis is Abaqus by DS Simulia.

Similar simulation and modeling software products are offered by companies such as MathWorks, which offers a suite of products called MATLAB and SIMULINK. These two tools are integrated into a product that is designed to build and simulate mathematical models, algorithm development, and many mathematical models (MathWorks, 2014). MATLAB and SIMULINK are development tools that are frequently used by software engineers in designing the battery management system (BMS) software, communication and controls systems.

COMSOL offers an analysis and modeling tool that is based on a "multiphysics" model which means that it can do thermal analysis, mechanical analysis, fluid flow analysis, electrical analysis, chemical analysis, and much more (COMSOL, 2014).

All of these tools and software packages represent some of the most common simulation and analysis tools that are used in the battery engineering and design process, but are by no means a complete list nor are they intended to promote one tool over the other. The best recommendation that I can make is that your engineering team must evaluate their engineering and evaluation needs, their skill sets and then discuss them with the software suppliers and select the right tool to meet their development targets and engineering budgets.

Analysis Tools

These different products, companies, and institutions integrate several different types of analysis into these tools. The most common type of analysis that is conducted for thermal systems is CFD. CFD is a set of tools for thermal analysis and heat transfer that can be used from the cell- to the system-level to analyze the impact of heat generation and cooling on all of the components in the system based on the "fluid" flow of either air of liquid. But it requires some programming and development by a qualified thermal engineer or analyst and when done will offer good thermal model that can be very closely correlated to the actual performance of a battery system. This, when used correctly, can be an extremely valuable tool for both evaluating various battery pack configurations and for speeding up the development process (Figure 1).

Another frequently used analytical tool is the finite element analysis (FEA) model. Where the CFD is focused on an analysis of the thermal performance the FEA model is a physics-based analysis tool that is more frequently focused on analyzing the mechanical forces and stresses that are placed on the cells, modules, packs, and system but can also be used for analysis of thermal and electromagnetic solutions. Using similar processes as the CFD analysis, a

Figure 1
Thermal models of lithium-ion cells.

representative model of the battery can be created and material properties can be applied. From this model different stresses and forces can be applied to the product in order to find failure modes and weak spots in the mechanical design.

Another tool that some thermal engineers use to evaluate the performance of energy storage systems is the lumped parameter model or lumped capacitance model. The lumped parameter model is a useful tool because instead of defining each and every characteristic and material type as is required in an FEA or CFD analysis, the lumped parameter model "lumps" the characteristics together into categories and makes the assumption that the differences within each of these groups is not important. The benefit of this model is that it can be very quickly developed and is quick to run and will provide directionally correct estimates of the thermal performance of the energy storage system. During the design process, it is very easy to continually make changes to your lumped parameter model in order to evaluate the impacts of those different changes on the thermal performance of the system. In this case, approximation will offer a good estimate of a system's performance while minimizing the number of complex calculations that must be made. One of the benefits of this model is that it can be developed using a simple Excel spreadsheet rather than needing to use a specialized software tool.

Hardware-in-the-loop (HIL) and software-in-the-loop (SIL) simulations are yet another set of tools that can help to quickly evaluate different performance profiles and different system designs in a safe laboratory-type environment. A HIL system uses a series of tools which could include some of the actual hardware but more likely the hardware will be represented by a mathematical simulation model. In the case of the battery pack, the hardware might include BMS controllers, electric motors, switches, and contactors or even a partial or complete battery pack. But for rapid turnaround most HIL engineers will create simulation models for many of the components. In this type of system, it is possible to test literally hundreds of different situations, failure modes, and performance models in a matter of hours. In the case of a battery controller, you may create simulations of the battery pack and then you can run each and every different fault code situation that has been developed by your software team to ensure that the system responds correctly and as it was designed.

As a final thought, it is important to note that none of these tools are intended for the typical sales person but instead a good thermal or mechanical engineer should be engaged to develop and use these tools due to the complexities involved.

Battery Sizing Tools

When it comes to lithium-ion battery sizing tools, there are not currently any industry standards developed in order to assist the system designer in generating an initial specification for a lithium-ion-based energy storage system. This is a weakness in the current literature on the subject. However, as mentioned previously several of the cell manufacturers have developed their own tools for doing this type of analysis and many others have developed tools that they use internally but do not distribute publicly.

It is possible to use some of the battery sizing models that IEEE organization has developed. IEEE standard number 485-2010 offers a set of guidelines for sizing a lead acid battery pack for stationary energy storage applications (IEEE, 2014) and IEE standard number 1115-2000 offers a similar methodology for sizing a nickel–cadmium battery for stationary applications (IEEE, 2014).

One battery sizing tool that is based on the IEEE standards is Alcad's "BaSiCs online Battery Sizing and Configuration System" (ALCAD, 2010). The software is free to download for a trial period. However, the tool was designed for lead acid type industrial applications which means that it will not always be a great fit for a lithium-ion application but with some manipulation and understanding of the standard it may do generally the same job for lithium-ion.

The biggest challenge with trying to adopt the lead acid sizing model to the lithium-ion battery application is the difference in load models. With the lead acid sizing model, it is typically possible to quickly add up all of the loads and times to determine the needed power. With a lithium-ion battery, the operation cycle tends to be much less static and will vary greatly throughout its operational profile. The lead acid models used in the IEEE standards are also very data-intensive when trying to convert them for lithium-ion use, requiring quite a lot of testing to be done in order to ensure the accuracy of the tool. There is also certainly quite a bit of lithium-ion testing that must be done, but for the sizing of the battery these data are typically part of the standard characterization process.

Design Guidelines and Best Practices

- Many analytical software packages are being modified and/or developed to assist the engineering team in the areas of thermal, mechanical, and electrical performance analysis
- Analytical engineering packages enable rapid and frequent design updates prior to hardware being built
- Computational fluid dynamics analytical software packages are used for thermal analysis of battery pack designs and systems
- Finite element analysis can be used for physics-based thermal, electromagnetic, and mechanical analysis, but is most frequently used in analyzing the stress and vibration of the mechanical designs
- No industry standard battery sizing tools currently exist for lithium-ion-based energy storage systems
- IEEE standards for lead acid and nickel-cadmium batteries may be able to be used as surrogates
- Many battery companies have developed their own battery sizing tools

CHAPTER 7

Lithium-Ion and Other Cell Chemistries

In this chapter, we will develop a basic understanding of the types of chemistries, form factors, and their benefits and challenges for many of the cells that are being used in energy storage systems (ESSs) today. We briefly review some of the other chemistries that are or have been used in various applications including lead acid, nickel metal hydride (NiMh), nickel metal chloride, sodium sulfur, sodium chloride and others. However, the main focus will be on the many different lithium-ion chemistries that are either in use today or are being developed for future applications.

Batteries are typically classified into one of two categories: either primary or secondary. The general difference is whether the batteries are rechargeable or not. Primary batteries are not rechargeable; they are single-use batteries and must be discarded after they have been discharged. Examples of these are the simple alkaline batteries that are used in many household devices. Secondary batteries are rechargeable, multiuse batteries that can be charged over again. The amount of life, or use, in a rechargeable battery is dependent on the chemistry and operating profile. Some cells can be charged and discharged hundreds of times; others can be charged and discharged thousands of times.

Before we begin to break down the definition of what a battery is, we should get a basic understanding of the different components that make up a battery cell. While this is somewhat simplified, there are essentially five main components that make up a battery. The cathode is the "positive" half of the battery cell, which is made up of a substrate of some sort that is coated with the active material. In lithium-ion batteries, the substrate is often a very thin film of aluminum. The anode is the "negative" half of the battery cell and is usually made up of a thin copper substrate that is coated with the active anode material. Between these two halves is a "separator" material that prevents the two halves from touching and creating a short circuit. These three components are assembled together to form the electrodes and are either wound or stacked to form what is referred to as a jellyroll. The fourth component of a lithium-ion battery is the enclosure, which is most often a can or pouch, in which the jellyroll is inserted. This may take the form of a metal can, a plastic housing, or a polymer type "pouch." Once this is done, the fifth element is added to the mix—an electrolyte. The electrolyte is the medium that allows the ions to pass back and forth within the cell. There are many other bits and pieces that may be included in a battery cell, such as a current interrupt device (CID) or a positive thermal coefficient (PTC) which is a resettable thermal fuse. But these are not included in all cell types or chemistries.

The next thing that we should do is to clearly define what a battery is and how it differs from other types of energy storage. Most simply, a battery is an electrochemical means of storing

The Handbook of Lithium-Ion Battery Pack Design. http://dx.doi.org/10.1016/B978-0-12-801456-1.00007-5

energy that operates by converting chemical energy into electrical energy. Linden and Reddy (2011, p. 1.3) define a battery as:

> *...a device that converts the chemical energy contained in its active materials directly into electric energy by means of an electrochemical oxidation-reduction (redox) reaction.*

Essentially what Linden and Reddy's (2011) definition means is that when a current is applied to the cell, a chemical reaction occurs between the anode and cathode inside the battery cell, whereby lithium-ions (anions) flow between the anode and cathode through the electrolyte thus generating an electrical current which can be converted into "work" (power) for various applications. The battery is unique in energy storage solutions in that it both generates the energy and stores it within the same device. This makes the electrical battery somewhat unique in the energy storage applications where virtually all other applications generate the energy somewhere else and the storage device does only that, stores the energy. For example, liquid gasoline and diesel, compressed natural gas, liquid propane gas, compressed hydro, compressed air, and other similar technologies are examples of energy that is created in one place and then stored in another and used to generate power in yet another place. In the internal combustion engine, for example, gasoline is stored in a tank, then is pumped into a fuel injection unit (or a carburetion device) where it is then burned to create energy (power) to move the cylinders.

Different battery types have different performance characteristics and are therefore appropriate for different types of applications. In Table 1, nonlithium-based chemistries, several of the most common nonlithium-based chemistry types are compared. The table is not all-inclusive but offers a good summary of some of the major nonlithium chemistries that are being used in ESSs. For example, traditional lead acid offers the shortest cycle life and lowest energy density but also offers the lowest cost. Secondary battery chemistries such as lead acid, nickel cadmium (NiCd), NiMh, sodium sulfur, and sodium nickel chloride are or were used in many of the early automotive electrification attempts. In fact, today NiMh is still the best selling battery chemistry for hybrid electric vehicles (HEVs). In the sections that follow, we will take a brief look at these chemistries in a bit more detail.

Lead Acid

Lead acid (PbA, LA, or LAB) is one of the oldest and most commonly used chemistries in energy storage applications today. Lead acid is a reliable power battery but suffers from low cycle life only achieving 300–500 cycles in the standard lead acid battery. Lead acid batteries take their name from the combination of lead plates that form the anode and cathodes and the sulfuric acid electrolyte in which they are immersed. Today lead acid is the standard battery used in engine starting, lighting and ignition (SLI) applications due to its high power capability. Early electric vehicles (EVs) also used lead acid as it was the most readily available and cheapest battery on the market. Lead acid is also still commonly used in backup stationary

Table 1: Non Lithium-based chemistries

	Lead Acid	Nickel Cadmium	Nickel Metal Hydride	Sodium Sulfur	Sodium Nickel Chloride
Chemistry descriptor	PbA/LAB	NiCd	NiMh	NaS	NaNiCl
Specific energy (Wh/kg)	30–40	40–60	30–80	90–110	100–120
Energy density (Wh/L)	60–70	50–150	140–300	345	160–190
Specific power (W/kg)	60–180	150	250–1000	150–160	150
Power density (W/L)	100	210	400	-	-
Nominal voltage (per cell) (V)	2.0	1.2	1.2	2.0	2.6
Cycle life	300–800	1000–2000	500–1500	1000–2500	1000
Self-discharge (% per month)	3–5%	20%	30%	0%	0%
Operating temperature range (°C)	−20 to +60	−40 to +60	−20 to +60	300 to 400	300 to 400
Cost (per kWh)	$150–$200	$400–$800	$200–$300	$350	$100–$300
Maintenance	3–6 months	30–60 days	60–90 days	None	None

power applications where weight and size are not issues and cost sensitivity is important. It is still the most readily available battery and continues to set the cost targets for other chemistries.

A lead acid battery consists of a grouping of lead plates or lead grids that are coated with a paste made of red lead and sulfuric acid. Those plates are then connected in series and immersed in an acid solution that acts as the electrolyte. The positive plates are typically made from lead dioxide while the negative plates are made from "spongy" lead. Many current plate or grid designs alloy a material such as antimony, tin, calcium, or selenium into the lead plate in order to improve the manufacturing process and quality and to add strength to the historically weak lead plate structures. The plates are most often separated by a plastic divider to separate the positive from the negative plates in order to prevent short circuiting. There are usually an odd number of plates and every other plate is connected together in series, one half form up the positive pole (cathode) and the remainder of the plates form the negative pole (anode) (Figure 1).

The active material in the plates (lead) absorbs sulfur from the acid electrolyte during discharge and then the sulfur flows back into the electrolyte when the battery is charged. Over time this process tends to cause the active material on the plates to flake off and collect at the

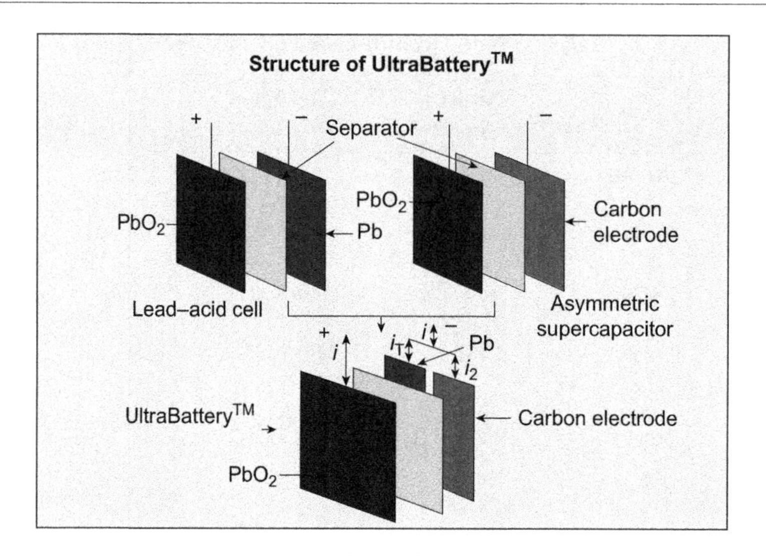

Figure 1
Schematic representation of UltraBattery™ configuration and operation. Soluble lead acid cell diagram, showing component materials. *Courtesy of CSIRO.*

bottom of the battery, this process is what limits the cycle life of traditional lead acid batteries. For this reason most lead acid batteries are designed with some additional space at the bottom of the cells for the flaked material to collect in order to prevent accidental short circuiting between the anodes. The other challenge with the traditional lead acid is the gassing that occurs as the acid reacts with the lead. This gassing is what drives the need for frequent testing of the electrolyte and the need to service and refill the battery.

Another type of lead acid battery that has been introduced to the market more recently is the absorbed glass mat (AGM) lead acid battery. The main difference between the traditional lead acid and the AGM is that in the AGM battery, a sponge type fiberglass mat separator is used to hold the electrolyte instead of submerging the lead plates in the sulfuric liquid electrolyte. This reduces the potential for leakage and spilling of the acid electrolytes. The AGM is therefore able to be "sealed" as the gasses are reabsorbed into the fiberglass mat. And as they are sealed, there is no need to service the batteries. The AGM battery falls into the category of being a valve-regulated lead acid (VRLA) battery. All batteries in this category are sealed and are able to recombine the gasses back into the electrolytes. This category of lead acid batteries are also frequently called sealed or maintenance-free for the same reason.

While they have the disadvantage of using hazardous materials such as lead and sulfuric acid, this is partially overcome as lead acid batteries are one of the most highly recycled products in the market today with a recycle rate in the United States of between 95 and 98%. This, of course, does not mitigate the environmental risk in regions of the world that do not recycle lead acid batteries to this same level. The other part of this equation that helps to drive high levels of recycling is that the used lead acid battery still has a residual value both to the end user and to the recycler. In this field, the used lead acid battery is most often resold back through a distribution center for a "core value," which is the value of the raw materials used in the battery that can be reclaimed through the recycling process.

Another one of the frequent arguments for lead acid batteries is that there are many standardized sizes available, they have become highly "commoditized." One method of categorizing the lead acid batteries was done by the Battery Council International (BCI) group numbers (Table 2).

While lead acid batteries are considered highly standardized and commoditized, the table below (Table 2) clearly shows that there are a lot of variations to this "commodity" product, many of which are extremely close in either size or performance to other standard form factors. This happens as new sizes are created that are specific to certain applications and to fit a wide range of products that span from small to very large. In addition to the dimensional characteristics that the BCI group numbers use to describe a lead acid battery, they also use voltage, terminal configuration, and position of the terminal to generate these categories.

It is this combination of high levels of commoditization, low material and manufacturing costs, and high residual value of the lead acid battery that has helped to drive it to the low costs that it currently experiences.

As noted above, the traditional lead acid and VRLA lead acid battery is most frequently used in automotive SLI batteries. However, they have also found frequent use more recently in applications such as data storage centers and backup power sources as well as in some industrial applications such as forklifts. Another application where lead acid is frequently seen is in those requiring extremely low costs, such as the e-bike market. Lead acid is a good fit for applications that requires standby type power, high power pulses, low cycle requirements, does not have weight or volume restrictions and requires very low costs.

While the traditional lead acid battery has changed little over the past 150+ years, there is some new work being done on advanced lead acid batteries that is making some great strides and is making lead acid solutions more appropriate solutions for some of the newer electrification projects such as stop/start type microhybrids and mild HEVs. New companies such as Energy Power Systems (EPS) and Ecoult have both been working on advanced PbA solutions. EPS has developed what they refer to as a planar layered matrix solution. EPS has taken an innovative approach to reimagining the lead acid battery and is achieving some extraordinary results claiming two to three times traditional lead acid cycle life as well as extremely high charge acceptance rates; in essence, they are achieving nearly NiMH performance at about one-third of the cost of NiMh. Targeting applications such as automotive microhybrids

Table 2: Passenger car and light commercial batteries

BCI Group Number	L	W	H	BCI Group Number	L	W	H	BCI Group Number	L	W	H
21	208	173	222	40R	277	175	175	61	192	162	225
22F	241	175	211	41	293	175	175	62	225	162	225
22HF	241	175	229	42	243	173	173	63	258	162	225
22NF	240	140	227	43	334	175	205	64	296	162	225
22R	229	175	211	45	240	140	227	65	306	190	192
24	260	173	225	46	273	173	229	70	208	179	196
24F	273	173	229	47	246	175	190	71	208	179	216
24H	260	173	238	48	306	175	192	72	230	179	210
24R	260	173	229	49	381	175	192	73	230	179	216
24T	260	173	248	50	343	127	254	74	260	184	222
25	230	175	225	51	238	129	223	75	230	179	196
26	208	173	197	51R	238	129	223	76	334	179	216
26R	208	173	197	52	186	147	210	78	260	179	196
27	306	173	225	53	330	119	210	85	230	173	203
27F	318	173	227	54	186	154	212	86	230	173	203
27H	298	173	235	55	218	154	212	90	246	175	175
29NF	330	140	227	56	254	154	212	91	280	175	175
33	338	173	238	57	205	183	177	92	317	175	175
34	260	173	200	58	255	183	177	93	354	175	175
34R	260	173	200	58R	255	183	177	95R	394	175	190
35	230	175	225	59	255	193	196	96R	242	173	175
36R	263	183	206	60	332	160	225	97R	252	175	190

and hybrids as well as stationary energy storage, the EPS technology offers a strong value proposition and appears ready to make big strides in the battery marketplace (Figure 2).

Australian-based Ecoult has also introduced another evolution of the traditional lead acid battery, which they have branded as the "UltraBattery." The UltraBattery has added a carbon electrode to the traditional lead acid battery, essentially adding an ultracapacitor into the lead acid battery. Designed to achieve more "partial state of charge" cycles than the traditional VRLA lead acid batteries, the UltraBattery has proven to be capable of achieving thousands of partial state of charge cycles with no serious degradation in performance or capacity and has been tested in applications such as renewable energy smoothing, grid ancillary support, and HEV automotive applications.

Both of these companies have proven that even after over 150 years, there is still more that can be done to improve the lead acid battery. They are proving to be very good and low-cost solutions to microhybrid vehicles, hybrid vehicles as well as for stationary grid storage and power backup systems.

If you are considering using a lead acid battery solution, there are a couple of trade and industry organizations that are dedicated to this product. One of the oldest and most important organizations in the lead acid industry is the BCI (Battery Council International, 2013) formed in 1925. BCI is a nonprofit trade organization working to bring together members of the lead acid community to develop industry standards, collect data, and generally promotes the exchange of information and ideas among its nearly 200 international members. Another important industry organization for lead acid batteries is the Advanced Lead Acid Battery Council (ALBAC) which was formed in 1992. The ALBAC is "...an international research and development consortium dedicated to enhancing the capabilities of the lead-acid battery to ensure its competitiveness in various energy storage markets" (Advanced Lead Acid Battery Consortium, 2011). While this is a somewhat smaller organization than BCI, it is an

Figure 2
Energy power systems planar layered matrix (PLM) battery.

important consortium of about 70 members whose goal is to enhance and engage with their member organizations in research programs dedicated to advancing lead acid batteries.

Nickel Metal-Based Chemistries

Another battery chemistry that has received a lot of attention in the 1990s and early 2000s is the nickel-based chemistries, including NiMh and NiCd. These chemistries achieved very high volumes with early portable power applications and subsequently, for NiMh, in HEVs. Nickel-based chemistries offered a higher voltage, higher capacity, and more cycles than traditional lead acid batteries but suffered from several challenges. Some of the challenges with nickel-based chemistries are the lower voltage and lower energy density compared to lithium-ion and NiMh suffers from "memory effect" that will reduce their available energy over time. The memory effect of these chemistries can be somewhat minimized by conducting periodic full discharges. Nickel-based chemistries also suffer from very high levels of self-discharge (Table 3).

Table 3: Nickel-based chemistries

	Nickel Metal Hydride	Nickel Cadmium	Nickel Zinc	Nickel Hydrogen
Chemistry descriptor	NiMh	NiCd	NiZn	NiH_2
Specific energy (Wh/kg)	30–80	40–60	70–110	50–65
Energy density (Wh/L)	140–300	50–150	130–350	55–110
Specific power (W/kg)	250–1000	150	280–2500	-
Power density (W/L)	400	210	420–7000	-
Nominal voltage (per cell) (V)	1.2	1.2	1.6	1.4
Cycle life	500–1500	1000–2000	300–900	>2000
Self-discharge (% per month)	30%	20%	20%	-
Operating temperature range (°C)	−20 to +60	−40 to +60	−20 to +50	-
Applications	Automotive hybrid electric vehicle (HEV)	Consumer electronics, power tools, light rail and train, uninterruptible power supplies, emergency lighting, telecom	Power tools, lawn and garden tools, light electric vehicles, HEVs	Satellite applications: Low earth orbit and geosynchronous earth orbit

Nickel Cadmium

Nickel cadmium (NiCd) batteries are most commonly used in portable power applications such as consumer electronics, portable power, telecommunications, uninterruptible power supplies and most frequently in train and rail car applications. In fact, about 40% of all NiCd batteries end up in one or another train-based application providing power from lighting to air conditioning to powering switchgear equipment and many other electrical power loads on the train (Erbacher, 2011).

The two biggest challenges, as mentioned above, that NiCd faces are that when compared to lithium-ion, it has a relatively low energy density (~40–60 Wh/kg) and the "memory effect" that NiCd faces. The memory effect occurs when a battery gets used to working within a certain duty cycle range. After time this will be the only range that the cell can operate within as it will modify its electrical properties to adjust to the new operational profile.

For these reasons NiCd has not found a large market in automotive application and is generally not appropriate for automotive electrification applications. However, NiCd has proven very suitable to consumer electronics and light rail type applications.

Nickel Metal Hydride

Nickel metal hydride (NiMh) gained initial market acceptance in portable power applications but really found a good home with the introduction of the automotive HEV. The HEV is a very good application for NiMh, as NiMh offers about twice the energy and power density of lead acid at about half the size. Typical NiMh cells will range between 30- and 80-Wh/kg-energy density.

Some early EV applications, such as the General Motors EV1, adopted NiMh batteries to power these EVs thereby offering significant improvements over the lead acid alternatives. However, with the introduction of lithium-ion chemistries with much higher energy density, NiMh became a less viable solution for full electrification. NiMh requires about twice the space as lithium-ion (higher volumetric energy density of lithium-ion) at about twice the weight (higher gravimetric energy density of lithium-ion). On the other hand, NiMh offers about half the size and weight of a comparative lead acid battery solution.

But NiMh has found a good home in the high-volume automotive HEV applications such as the Toyota Prius which now has well over 7 million HEV vehicles on the road globally with this technology since the launch of their first generation Prius (Holmes, 2014). Additionally, with the growth of stationary power and telecom power applications, NiMh offers many benefits over what has traditionally been a lead acid application market. Yet as the cost of lithium-ion comes down more companies are beginning to evaluate other chemistries.

Other Nickel-based chemistries, such as nickel zinc (NiZn) and nickel hydrogen (NiH_2) have found use in many applications. NiZn has been integrated into applications such as power tools, lawn and garden equipment, light electric vehicles (LEVs), consumer cells (AA, AAA, etc.), and even in some HEV demonstration applications. NiZn chemistries are relatively abuse tolerant with high discharge capabilities and are relatively low cost. More importantly, NiZn are more environmentally friendly than the NiCd or even NiMh cells that are available today. NiH_2 has found a very good home in aerospace satellite applications for both low earth orbit (LEO) and geosynchronous earth orbit (GEO) satellites. While NiH_2 has a higher initial cost and relatively low energy density, it is also capable of surviving long cycle life and calendar life while in orbital applications. LEO satellite applications typically require over 35,000 cycles and calendar life of over 6 years. GEO satellites require somewhat shorter cycle life, but still need about 2000 cycles but must survive much longer calendar life, up to 20 years in some applications (Brill, 2011).

Various other nickel-based chemistries have been tested and evaluated over the years, however, these four represent the most commonly used nickel-based chemistries that are available in the market today.

Sodium-Based Chemistries

Another set of chemistries that has gained recent interest are the sodium-based batteries. Sodium batteries are often referred to as thermal batteries due to the fact that they must operate at very high temperatures in order to maintain the sodium (salt) electrolytes in a molten (liquid) state. One of the earliest developments was by a research team in South Africa called the Zeolite Battery Research Africa Project (ZEBRA). The work of this group eventually became the basis for most of the current sodium batteries on the market. Companies such as ZEBRA technologies, General Electric (GE), SONICK-FIAMM (MES-DEA) are producing ZEBRA-type sodium metal halide batteries (Table 4).

One of the interesting characteristics of the sodium chemistries is that they need to operate at very high temperatures, typically between 350 °C–700 °C. This means that they operate in a molten form, which in itself creates some challenges. In the typical molten sodium battery, the anode and cathode are separated by a solid ceramic separator. In these batteries, the electrolyte is typically a molten sodium of either aluminum chloride, nickel chloride, or similar chemistry. The positive electrode is made of nickel and the negative is the molten sodium. The ceramic separator separates the anode from the cathode, but is only activated once the sodium reaches its melting point, which may be as low as 150 °C but is typically closer to around 350 °C.

These chemistries are very attractive because of the inherently low cost of the materials involved and for the ZEBRA-type sodium nickel–metal halide battery the inherently safe failure mode. The NaS batteries are more energetic if the separator fails. One of the challenges with this chemistry is that if it is allowed to cool below about 160 °C, the sodium will

Table 4: Sodium-based chemistries

	Sodium Metal Halide	Sodium Sulfur
Chemistry descriptor	Sodium aluminum chloride ($Na-AlCl_4$) Sodium nickel chloride ($Na-NiCl_2$)	NaS
Specific energy (Wh/kg)	90–120	110
Energy density (Wh/L)	160	-
Specific power (W/kg)	150–180	150
Power density (W/L)	–	-
Nominal voltage (per cell) (V)	2.6	2.1
Cycle life	1000–1500	1000
Activation temperature range (°C)	270–350	350–700
Companies	ZEBRA, Eagle-Picher, GE, SONICK-FIAMM	
Applications	Automotive electric vehicle, military, space, telecommunications, train and rail, and stationary energy storage	Space and satellite applications

solidify and heating it back up is a time-consuming process that can take 12 hours or more. Another challenge is that this battery cell generally requires significant protection due to the high operating temperatures, often requiring a well-insulated module and pack to protect operators and users from the high temperatures.

One of the biggest benefits of sodium chemistries is cost, as the cathode is based on sodium, aluminum, nickel, and sulfur, all very common materials, it tends to be a very cost-effective solution. Another benefit of the sodium chemistry is that, compared to lead acid, it offers both higher energy density and higher power density. With energy and power density levels approaching that of both NiMh and lithium-ion phosphate chemistries, sodium begins to become a competitive solution.

Development work continues on these chemistries but so far applications have been limited to several automotive demonstrations and some stationary energy storage applications. In the stationary energy storage field, the sodium-based thermal batteries are making some very good in-roads due to the low cost and the relative isolation of the ESS from the workers.

Lithium-Ion Cells

The first commercial lithium-ion chemistry was introduced to the market in 1990 based largely on the work of Dr John Goodenough of the University of Texas. From its introduction in 1991 to the early 2000s, sales of lithium-ion grew in demand to become the highest volume cell manufactured in the world with about 660 million small cylindrical cells and another 700 million small polymer (pouch)-type cells manufactured annually in 2013 (TrendForce, 2013).

Lithium-ion quickly became the battery of choice for most small electronics because it contained much higher energy density than comparable cells on the market. This meant that you could create a battery with the same energy as NiMh but it would be about half the size and half the weight. For portable power applications such as laptops and cell phones, this meant longer run times and longer life batteries.

How does lithium-ion work? Well, in fact, it works much like any other battery type, as a current is applied ions are forced from the anode to the cathode. In the lithium-ion battery, energy flow is created as the lithium-ions within the cathode are transferred through an electrolyte medium into the anode, this represents a charging event. A discharging event is represented by the lithium-ions being transferring through an electrolyte medium from anode into the cathode (Figure 3 below). This does seem counterintuitive for most of us but when the battery is discharging the ions *are* passing from anode to cathode. In the diagram below a charging event is being shown, with the lithium-ions passing from the cathode material through the electrolyte to the separator and then again through the electrolyte and to the anode material. This action creates a voltage flow up the copper current collector and, in a closed circuit loop, to the positive current collector.

One of the other benefits of lithium-ion chemistries over nickel- and lead-based batteries is the higher voltage. Typical NiMh and NiCd rechargeable cells operate at about 1.2–1.5 V nominal, whereas lithium-ion cells typically operate between 3.2 and 3.8 V nominal. Having a higher voltage is important in that it means that you need to connect fewer cells together in series in order to achieve your desired pack voltage. For instance, a NiMh battery pack with 350 V may require 292 cells to achieve that voltage (350 V/1.2 V = 292 cells). Whereas a lithium-ion-based battery pack would only require about 98 cells to achieve the same system voltage (350 V/3.6 V = 98 cells).

In addition to having higher voltage and energy density, lithium-ion also has a lower rate of self-discharge. This means that its natural capacity loss over time when the batteries are in

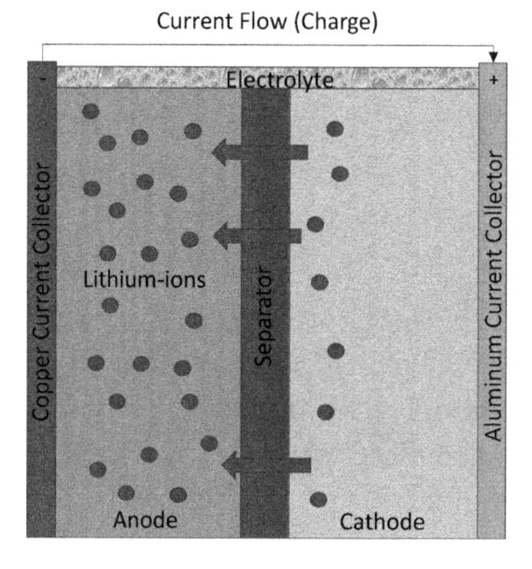

Figure 3
Lithium-ion cell ion flow.

storage is less than that of other chemistries with many lithium-ion chemistries losing only 1–5% per month. Capacity loss during storage comes in two types, reversible and permanent. Reversible capacity loss is that energy that is lost during storage but will be regained once the battery is cycled again. Permanent loss is that portion that will never be regained again. Virtually all lithium-ion chemistries have *some* amount of reversible capacity loss over time, some of which is generally always permanent.

Finally, lithium-ion chemistries tend to have much better cycle life than the other chemistries. Where PbA may only get 300–500 cycles before it reaches its end of life (EOL), lithium-ion can achieve thousands of full discharge cycles before reaching its EOL during 100% depth of discharge (DOD) cycles. If we look at partial cycles, the lithium-ion battery will be able to achieve tens of thousands of cycles as the DOD is reduced. For example, if we look at a typical lithium-ion chemistry, it may achieve 1000 cycles using 100% DOD, but if we take that same cell and use only 80% of its total usable energy, we will find that we can now get several thousand cycles.

Table 5 below summarizes some of the general performance characteristics of some of the most common lithium-ion chemistries that are in use today, including nickel manganese cobalt (NMC), nickel cobalt aluminum (NCA), lithium iron phosphate (LFP), lithium titanate (LTO), lithium manganese oxide (LMO), and lithium cobalt oxide (LCO).

Table 5: Lithium-ion chemistries

	Lithium Iron Phosphate	Lithium Manganese Oxide	Lithium Titanate	Lithium Cobalt Oxide	Lithium Nickel Cobalt Aluminum	Lithium Nickel Manganese Cobalt
Cathode chemistry descriptor	LFP	LMO	LTO	LCO	NCA	NMC
Specific energy (Wh/kg)	80-130	105-120	70	120-150	80-220	140-180
Energy density (Wh/L)	220-250	250-265	130	250-450	210-600	325
Specific power (W/kg)	1400-2400	1000	750	600	1500-1900	500-3000
Power density (W/L)	4500	2000	1400	1200-3000	4000-5000	6500
Volts (per cell) (V)	3.2-3.3	3.8	2.2-2.3	3.6-3.8	3.6	3.6-3.7
Cycle life	1000-2000	>500	>4000	>700	>1000	1000-4000
Self-discharge (% per month)	<1%	5%	2-10%	1-5%	2-10%	1%
Cost (per kWh)	$400-$1200	$400-$900	$600-$2000	$250-$450	$600-$1000	$500-$900
Operating temperature range (°C)	−20 to +60	−20 to +60	−40 to +55	−20 to +60	−20 to +60	−20 to +55

Lithium-ion batteries have relatively few components in them, in fact there are really only about five major components and a typical bill of materials for a cell may have only between 10 and 20 line items (Figure 4). Regardless of form factor, the major components of the cell are the cathode, an aluminum foil in many designs that is coated with the active cathode material. The second component is the anode, a copper foil that is coated with the active material of the anode. These two coated foils are kept separate through the use of a separator material most often made of some type of polypropylene (PP) or polyethylene (PE) plastic. This forms the jellyroll, which is then inserted into the container or housing. This can be either a metal can, a plastic enclosure, or a metal foil-type pouch. Into this assembly is then injected the electrolytic liquid. The entire assembly is then hermetically sealed and is ready to move into the next stage of cell manufacturing. Figure 5 shows a high level descriptor of the main components of a lithium-ion cell in relation to its cost.

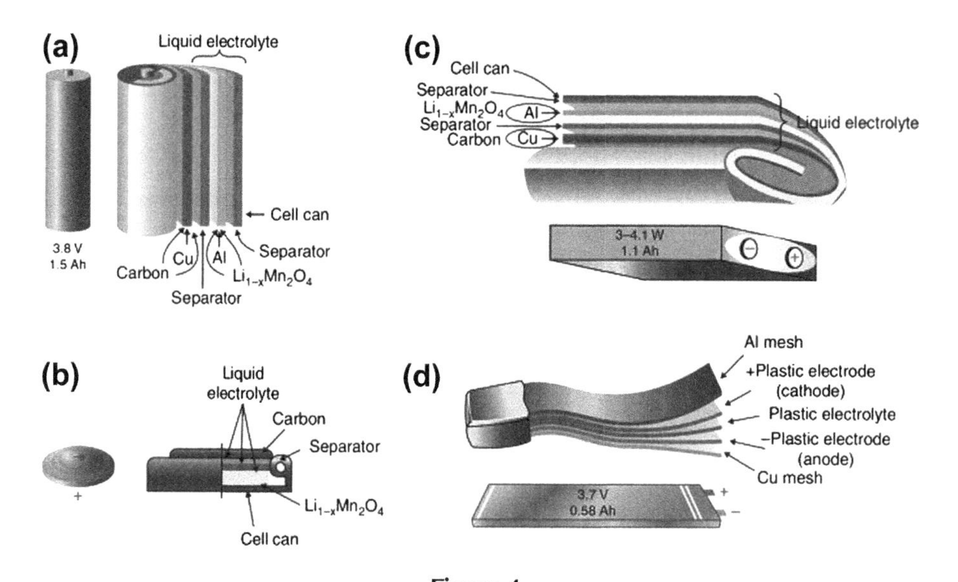

Figure 4
Prismatic lithium-ion cell components.
Source: © Elsevier, Encyclopedia of Electrochemical Power Sources, P. Kurzweil, Lithium Rechargeable Systems, vol. 5.

Cathode Chemistries

Lithium-ion is a very general term that refers to a wide variety of different chemistries, each of which has very different performance characteristics. The most common chemistry combinations used in lithium-ion cells include: LFP, NMC, LCO, NCA, and LMO. And of these five major chemistries, each cell manufacturer may use them in different combinations to achieve different performance results or in some cases the cell manufacturer may actually combine different chemistries in order to get the different benefits of each chemistry into one-cell design.

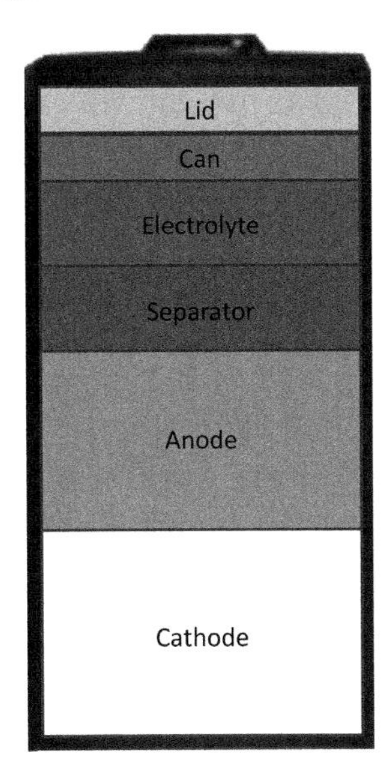

Figure 5
Lithium-ion cell components.

LFP is one of the most common chemistries in automotive applications due to its having a high power capability and relatively low cost. This means that it can accept a regenerative braking charge and can provide an acceleration discharge very quickly. The other reason that LFP is frequently used is due to its relatively low cost. In relation to some of the other more rare materials used in lithium-ion cells, iron phosphate is quite common and low cost. Another reason that LFP has gained a high level of usage is that is has been deemed a "safer" chemistry than others. However, this is somewhat of an inaccuracy. All lithium-ion chemistries have similar failure modes; it is only a matter of at what temperature they fail. In this case, as LFP has lower energy density than the other chemistries on the market and that means that there is less energy to discharge in the event of a failure. However, it is fair to say that LFP is more tolerant of abusive conditions such as overcharging the cell and high temperatures.

With the introduction of production EVs, lithium NMC chemistries have begun to gain a strong foothold due to its higher energy density and higher voltage. Depending on the combination of materials, it is also sometimes referred to as nickel cobalt manganese (NCM) if there is a higher percentage of cobalt than manganese in the chemistry. NMC shows a relatively high nominal voltage of about 3.6–3.8 V per cell and has a one of the highest energy densities in a production cell today of between 140 and 180 Wh/kg in production applications with some chemistries exceeding 200 Wh/kg.

LCO is most commonly used in handheld electronics such as cell phones, cameras, and laptop computers. While LCO offers generally higher energy density and long cycle life, it suffers from being less stable at higher temperatures and more reactive than other chemistries. This means that above about 130 °C the cell will enter the thermal runaway stage, a lower temperature than other lithium-ion chemistries. For this reason, LCO has not seen wide use in large application but continues to see use in small consumer electronics and in fact has been outright rejected by some automotive manufacturers. LCO is also a somewhat higher cost chemistry due to the high amount of cobalt, a relatively precious material that it contains.

NCA is frequently used in portable power applications but is being evaluated for automotive applications due to its high power, however, it does have some drawbacks that are keeping it from gaining a foothold in the vehicle space—high cost and low safety.

Finally, LMO offers high energy and high power, however, suffers from shorter cycle life thus making it an appropriate chemistry to be used in portable power applications where you want long run time but not necessarily in automotive applications where you want long life.

Anode Materials

The negative side of the electrode is known as the anode. Today, most anodes are made of a mixture of one of two materials either graphite, or soft or hard carbons. There are many different grades and types of graphite, with graphene forming the basis for most of it. But the quality and selection of the material will play a very important role in the performance of the cells (Dahn & Ehrlich, 2011).

One other lithium-ion anode that is gaining a lot of interest is the lithium-ion titanate cell. An LTO cell offers the advantage that it can operate at a lower temperature than other chemistries and offers a high power density. However, LTO suffers from having a lower nominal voltage, in the range of about 2.2–2.3 V per cell. In addition, current LTO offerings tend to be much higher cost than comparative NMC or LFP cells. This means that not only do you need more of them to achieve the desired system voltage, they will also cost more. However, there are several manufacturers that are working on improving the cost of LTO which may make it a good fit for many low energy and high power applications such as micro-HEVs and some are working hard to integrate LTO into fully EVs. There does not appear to be any technological reason that LTO will remain as a high-cost solution, as volumes begin to increase the costs should also decrease at a relatively stable rate.

Much research is being done into several different new anodes including silicon, tin, germanium, carbon nanotubes, and other nanocomposite materials. However, most of these are still in the laboratory and none are currently commercialized for high-volume products. As can be seen in Figure 6 below, there are some very big opportunities to increase energy density by moving to silicon, tin, or some of these other types of anode materials. There are still some

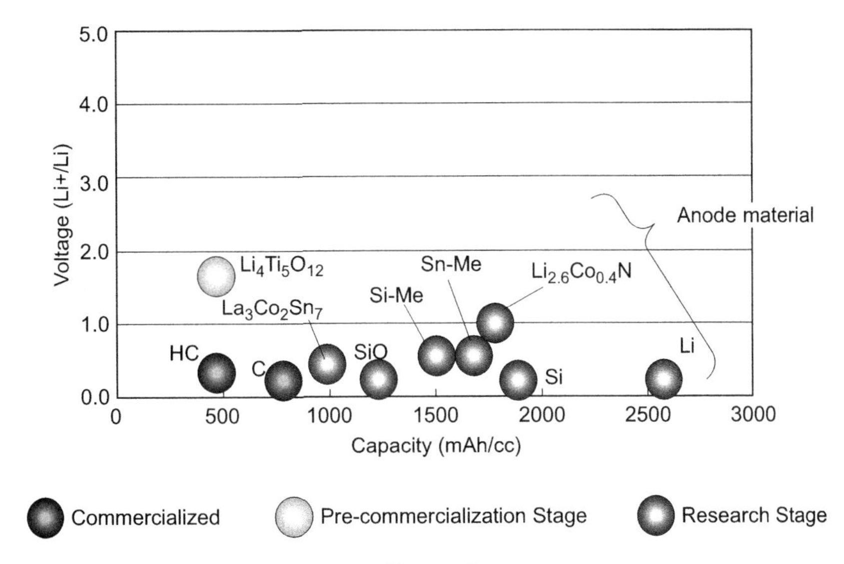

Figure 6
Anode material performance comparison.

major hurdles that need to be addressed before they can be commercialized. For instance, both tin and silicon cause a volumetric growth of between 250 and 300% during cycling which causes reduced cycle life. Today, these are all still in the early development stages but there is good hope that in the not-too-distant future we will be able to overcome some of the obstacles to make these anode chemistries realities.

Separators

The next component within the lithium-ion cell for us to consider is the separator. A separator is a thin piece of material, often a plastic or ceramic, which separates the anode from the cathode. In use, the separator material must be able to withstand the often corrosive hydrocarbon (HC)-based electrolytes that are used in lithium-ion cells while still maintaining the isolation of the two electrodes within the cell. The main purpose of the separator is to do just that, to separate the anode from the cathode. If the two halves of the electrode come into contact, an internal short circuit will occur, causing a cell failure. Therefore, the separator is of key importance in any lithium-ion cell design.

Some cells use a multilayer PP or PE plastic to separate the anodes and cathode. This material allows the lithium-ions to pass from anode to cathode while still protecting the cell from short circuiting. Some cell manufacturers use a trilayer PP/PE/PP in order to allow the middle layer to melt at higher temperatures (~135 °C) while still ensuring the separation of anode and cathode by ensuring that the outside PP layers do not melt until they reach a higher temperature (~155 °C). At high temperatures, the pores in the PP begin to melt which prohibits the flow of lithium-ions. However, for many separators the plastic begins to shrink and entirely melt at somewhat lower temperatures causing an internal short circuit in the cell and thereby permanent failure above temperatures of about 90°–110 °C.

Some lithium-ion cell manufacturers are beginning to integrate ceramic-layered separators in their cells as it enables higher temperatures and increases the safety of the cell, especially during nail penetration testing events. This also tends to improve the power rate capability of the cells by reducing the internal resistance.

Electrolytes

The electrolyte is usually a liquid- or gel-based solution that the anode and cathodes are emerged in which acts as a conductor allowing the lithium-ion cells to move between the anode and the cathode. The electrolyte is typically a HC-based mixture that includes multiple additives which provide different functionality within the cell. The use of additives is the cell manufacturer's "secret sauce" and is one of the key areas of intellectual property of the cell makers. Typical electrolytes may include a mixture of alkyl carbonates such as ethylene carbonate, dimethyl, diethyl, and ethylmethyl carbonates and lithium salts (LiPF6) (Aurbach, et al., 2004).

These additives offer various different benefits during operation. The main functions of the additives are to facilitate the formation of the solid electrolyte interphase (SEI) layer, to reduce the irreversible capacity and gas generation of the SEI formation and long-term cycling, and to enhance the thermal stability of the cell (Zhang, 2006). For example, in cylindrical- and can-type cells, an additive may enable "gassing" or gas generation at pre-defined temperatures in order to engage the CID, a resettable fuse internal to the cell.

SEI layer formation is another important function performed by the electrolyte. As the first electrochemical reactions begin to occur inside the cell, the electrolyte begins to react with the graphite or carbon anodes. Some of the lithium begins to form a passivation layer on the surface of the anode; this is referred to as the SEI layer (Dahn & Ehrlich, 2011). It results in an irreversible loss of capacity during the formation of the cells and generally during the very first cycle. However, this nonactive layer of lithium now begins to protect the graphite or carbon surface from further chemical reactions with the electrolyte.

One of the challenges associated with the electrolyte is during the failure and cell venting. As an HC-based liquid, the electrolyte will burn when ejected from the cell during a thermal runaway event. Due to this, there has been much interest in both gel electrolytes and solid electrolytes as well as in aqueous-based electrolyte solutions that are nonflammable.

Gel electrolytes have many of the same features as a liquid but as they are a gel-type material they have less potential to "flame" up during a failure event. Solid electrolytes are being evaluated by many manufacturers as well, but most are not yet ready for high-volume production. The solid electrolytes are generally being deposited onto either a separator layer or directly onto the cathode material.

Safety Features

Like many other technologies, lithium-ion cells have some inherent safety challenges. In essence, you have a container of energy that under certain conditions can be released. The amount of energy released will depend on the size (capacity) of the lithium-ion cell. Additionally, there is always the potential for contamination or errors to occur during the cell manufacturing process, which could result in an early failure or energy release.

In order to help mitigate these risks, there are several safety features that many cell manufacturers have integrated into their designs. The first and most common among a wide variety of cell types is the vent. The cell vent is basically designed as an engineered failure point within the cell in the event of a buildup of pressure inside the cell. The vent is designed to open up in the event that there is a pressure buildup within the cell. An engineered vent is generally only included in cells that have a hard can-type container; pouch-type cells use a slightly different variation of the vent. The pouch cell is more likely to be designed with a notch or other point in order to attempt to direct the failure in a specific area. Without this notch, the pouch type cell is much more difficult to forecast where the cell will fail.

The CID is a type of nonresettable fuse that is often integrated into the internal cell components. The CID is an engineered fuse that is most often pressure-based. So the CID is integrated into hard can-type cells as it is much more difficult to use a pressure-based fuse in a pouch-type cell. In essence, the CID is a two-part mechanism that is designed to separate and break the flow of current to the terminals if the pressure builds up beyond a certain point.

Another type of safety feature that is included in some cell designs is a thermal fuse, referred to as a PTC device. The PTC is essentially a resettable thermal fuse that is designed to open if the temperature cell rises above a predetermined point. If this PTC opens, the current flow between the cell terminals and the jellyroll is temporarily broken. If the temperature of the cell falls back down to normal operating temperatures, the PTC will reset and the cell will again be usable. The challenge with the PTC is that it will generally limit the maximum voltage that can flow from a cell and is therefore limited to small cells in small portable-type applications, typically below 26 V or so.

The other safety feature is one that we have already actually talked about, the separator. In lithium-ion cells, the three types of separators PP, PE, and ceramic-coated separators also perform a safety function. In the plastic-based (PP, PE) separators, they are engineered to fail (melt) at a predefined temperature. In fact, some cells use a three-layer PP/PE/PP cell which allows the center layer to melt at a temperature about 20 °C lower than the PP layers. This prevents the lithium-ions from flowing between anode and cathode and is intended to slow/

prevent a failure event. In the ceramic-coated separators, the ceramics have a much higher temperature at which they will fail and so may enable higher cell temperatures before thermal runaway onset. Two other benefits of ceramic separators is that they will not shrink like the PP or PE separators do and the ceramic separator is also believed to help improve the ways the cells fail during nail penetration events.

These safety features can be designed into many of the cell types, but the traditional 18650 cylindrical cell most often has the CID- and PTC-type safety feature. In the larger cell sizes, these features are a bit more difficult to integrate and as noted above the PTC will limit the maximum voltage so is not appropriate for cells that are intended for high-voltage systems.

Lithium-Ion Cell Types and Sizes

There are essentially three main types of lithium-ion cell form factors: small cylindrical, large prismatic, and pouch (or polymer) cells. By far the highest volume lithium-ion cell format in production today is the 18650 cylindrical cell with nearly 660 million cells produced annually (TrendForce, 2013). The nomenclature 18650 means that the cell is 18 mm diameter by 65 mm in length. When it comes to automotive applications, nearly all of the major auto manufacturers have identified "small" cylindrical cells as being appropriate mainly for u-HEV- and HEV-type power applications, with the exception of Tesla who is using exclusively high-volume 18650 cells. For plug-in hybrid electric vehicle (PHEV)- and battery electric vehicle (BEV)-type applications, the majority of the auto manufacturers are using either large rectangular or cylindrical prismatic cells or flat "pouch"-type cells. The reason for this is mainly due to the quantity of cells required to achieve the voltage and energy needed which from a reliability analysis means that there are many more connections with small cells than when using a larger cell and therefore many more potential areas of failures in assemblies of small cells. From the Original Equipment Manufacturer (OEM) perspective a high quantity of cells equates to a high potential for future warranty cost. They tend to look at this from a quality perspective: with over 7000 cells that means over 14,000 connections to those cells, compared with 576 welds for the volt pack. So from a "Six-Sigma" quality perspective, there is a much higher percentage chance that one of the 14,000 connections will fail than in a pack with only 576 cells connections.

As an example, the Tesla uses an 18650 style cell from Panasonic and in their Model-S Sedan, they require over 7000 cells in their 85-kWh-pack (they also offer a 60-kWh-"base" pack). While the General Motors uses only 288 pouch-type cells in their 16.5-kWh-Chevrolet Volt and Nissan only uses 192 pouch-type cells in their 24-kWh-LEAF.

As we begin looking at physical types of cells, there are basically three cell types that are in mass production today: cylindrical, prismatic, and polymer. As has been stated above, the 18650 is by far the highest volume cell in production today. However, there are many other small cylindrical cells being produced from a 32330 (32 mm diameter × 330 mm length) produced by A123 and the 18 mm by 36 mm by 65 mm (essentially the same size as two

18650 cells side by side) “Swing” and “Sonata” cells produced by Boston-Power. Saft produces a series of large cylindrical cells that are in wide use in aviation, aerospace, and military applications. The benefit of the cylindrical cell is that it uses a steel, nickel-coated steel, or aluminum can which offers a high-strength packaging requiring a lot of energy to damage it and it provides “stack” pressure on the jellyroll inside the can. There are also a significant number of cell assembly and manufacturing equipment suppliers that provide “off-the-shelf” manufacturing solutions for this product today.

One of the challenges involved in using smaller format cells, aside from size, is in the lid assembly. Many manufacturers have transitioned to laser welding the lids onto the cells, while some manufacturers continue to use a crimping process to attach the lid. The concern over crimping was exemplified in 2006 with the $250 million dollar recall issued by Sony when cells started to fail when metallic particles began to make their way inside the cells (Mook, 2006). Subsequent research found that it was during the crimping process some of the nickel that was used to plate the steel cells was flaking off and causing the internal short circuits and fires in consumer applications.

Another disadvantage, yet a relatively small one, is that the cylindrical cells tend to have much higher initial impedance than a comparative prismatic- or polymer-type cell. This means that it will have a higher rate of heat generation and in this format the best cooling solution is to air-cool the pack. However, this is also the least effective active cooling methodology. Yet, because the cells are small, the heat generation is easily manageable.

Prismatic-type cells, a hard steel, plastic or aluminum can in a rectangular shape, is quickly gaining a lot of attention from major auto manufacturers. One of the reasons for this is that the Verband der Automobilindustrie (VDA), an auto industry association made of European auto makers, has brought forth one of the first sets of proposed “standards” around cell sizing (see Table 6 below). The benefit of the larger prismatic cell is that it requires less “pack

Table 6: Verband der Automobilindustrie (VDA) standard prismatic cell formats

	Standard VDA Cell Sizes					
	HEV	PHEV1	PHEV2	PHEV2+	EV1	EV2
Form factor	Prismatic can					
Capacity (Ah)	5.5–7	20–25	25–32	35–37	40–50	60+
Length (mm)	120	173	148	148	173	173
Height (mm)	85	85	91	125	115	115
Thickness (mm)	12.5	21	26.5	26.5	32	45
Volume (L)	0.128	0.309	0.357	0.490	0.637	0.895
Ah/L	55	65–81	62–76	65–76	63	67
Wh/L	155–200	234–292	255–275	260–275	226	241

EV, electric vehicle; HEV, hybrid electric vehicle; PHEV, plug-in hybrid electric vehicle.

hardware" to integrate into the complete ESS when compared to a polymer-type cell and they come in quite high capacity ratings. And as mentioned earlier, there are fewer cell-to-cell connections that need to be made so the reliability is expected to be higher. I say expected to here since there are few of these applications that have reached their EOL yet, so the reliability calculations are still estimates for most of them.

The third format for lithium-ion cells is the lithium-ion polymer cells (LiPo, LIP, or Li-poly), also called "pouch" or "laminate" cells because they use a soft polymer laminate casing. The term polymer was originally used in reference to the use of a polymer-based electrolyte instead of a liquid-based electrolyte. However, today the term has become a general moniker used to refer to the physical form factor of the pouch-type cell. This does on occasion cause some confusion when discussing lithium polymer cells as some people refer to them in the polymer electrolyte aspect and others in the form factor.

The pouch type cell has become the standard in many of the portable power applications such as cell phones and tablets due to their thin form factor. It is also a very flexible cell-type solution because it is relatively easy to create many size variations of this cell, making it relatively easier to design into unique pack solutions. One of the challenges of this form factor is that it is more difficult to integrate some of the safety features mentioned earlier. Cell venting is much more difficult to predict and manage, and it is generally not possible to integrate things like CIDs and PTCs into this form factor. It is also generally necessary to design the cells into modules that can manage the "stack pressure" of the cells. In most cases, the pouch-type cell will perform better over life, if a consistent pressure is applied over the face of the cell. It is also very important to ensure that the pressure is uniformly applied. If the pressure is not applied uniformly, it can affect the cell's ability to pass the lithium-ions back and forth within the cell and eventually cause them to begin getting stuck. This is known as lithium plating, or in some areas it is referred to as "white out" when the lithium-ions become fixed and no longer pass back and forth; this increases the impedance of the cell and reduces the cell life. The other challenge of the polymer-type cell is that it must be handled with some care because the packaging is a soft-type laminate or polymer enclosure. So it is relatively easy to damage during cell assembly. Cells in this size can range from milliampere hours (mAh) all the way up to about 100Ah in capacity.

Lithium-Ion Cell Manufacturers

Today, there are hundreds of lithium-ion cell manufacturers in the marketplace. In this section, I will briefly review what are generally considered to be some of the top players in this market. This is only intended to be a short introduction to a handful of manufacturers, there are plenty more out there that I will not cover, so do not assume this is an all-inclusive list. I will leave some of the start-ups with innovative chemistries for the final chapter.

A123/Wanxiang—A123 was a start-up based in Waltham, Massachusetts and founded as a spin-off from a group of MIT researchers. A123 made some very significant early progress with their advanced LFP, chemistry they dubbed "nanophosphate." A123 did some of the original design work for the Chevrolet Volt. They were also a recipient of the U.S. Department of Energy (DOE) funding. A123 installed large stationary systems, automotive systems, and military systems among many others. However, after a forced recall of the Fisker battery system, A123 ended up filing bankruptcy. They successfully came out of Chapter 11 filing with Chinese company Wanxiang as the new owners. A123/ Wanxiang continues to operate today selling cells and systems. Their headquarters are in Massachusetts, cell and pack manufacturing and engineering in Michigan, and cell manufacturing in China.

ATL—Amperex Technology Limited (ATL) and their subsidiary Contemporary Amperex Technology Limited (CATL) are Chinese-based cell and pack manufacturers. As one of the largest producers of pouch-type lithium cells for consumer electronics, ATL has grown into a major producer of large format prismatic cells for grid, stationary, automotive, and other large applications.

Bosch/Lithium Energy Japan (LEJ)—German-based Bosch has been developing battery pack solutions for years. They had at one point a joint venture with Samsung (SB LiMotive) to offer a battery pack for the Fiat 500E. After the breakup of the joint venture (JV) with Samsung, Bosch entered into a relationship with Lithium Energy Japan (LEJ) in much the same structure as the original SB LiMotive JV.

Boston-Power—Massachusetts-based Boston-Power developed a unique lithium-ion form factor, with a cell that is the same size of two 18650 cylindrical cans side by side. Their headquarters are in Westborough, Massachusetts and manufacturing is based in both Taiwan and China. Their initial market focus was on portable power products such as laptop computer batteries, but they have grown into industrial- and automotive-type applications.

BYD—Starting as one of the largest lithium-ion cell manufacturers in the world Chinese-based BYD has developed an LFP chemistry and moved from being a cell manufacturer to begin a fully integrated vehicle manufacturer. They offer several HEV, PHEV, and BEV vehicles in China and are moving into the HEV and electric bus business.

Continental/SK Energy—German-based Continental and Korean-based SK Energy formed a JV building on the expertise of both companies. Continental has been working on engineering and battery system design for over 10 years. SK Energy is a relative new player in the lithium-ion cell market with an NMC chemistry and operations in Germany, Korea, and the United States.

Electrovaya—Electrovaya is a Canadian-based company with global headquarters in Mississauga, Ontario and its US headquarter is in Saratoga County, New York. The company was originally formed in 1996 and became a public company in 2000. Electrovaya focused on pouch-type cells for various applications from automotive to stationary and space-based applications.

EnerDel—a subsidiary of Ener1, EnerDel was formed as an outcome of a partnership between Ener1 and Delphi (Ener-Del). With an NMC-based pouch cell, EnerDel has their

headquarters in Indianapolis, Indiana and have many applications in military, grid and stationary, and automotive applications.
Deutsche Accumotive—Deutsche Accumotive is a German-based manufacturer that is partnered with Daimler AG for the development of pouch-type cells for automotive applications.
Johnson Controls Inc (JCI)—Johnson Controls initially formed a JV with Saft, called Saft-JCI. Later the JV broke up and both companies went their separate ways. JCI focused on the VDA format cell with an NMC chemistry. JCI's target has been with the automotive applications. JCI-Saft was also one of the recipients of the U.S. DOE funding; after the breakup of the JV, JCI continued their development with operations in the United States.
LG Chem—LG Chem partnered with the US-based Compact Power (CPI) to build their market in the US market. LG has quickly become one of the largest suppliers of pouch-based NMC cells for automotive applications in the world. With the backing of the Korean conglomerate LG, they have focused on cell, module, and pack development throughout the world. LG Chem was also one of the recipients of the U.S. DOE funding.
Lishen—Tianjin Lishen is a Chinese-based lithium-ion battery manufacturer who has been building 18650-based cells and packs focused in the Chinese market since 1997. Today Lishen offers a variety of lithium-ion cells as well as ultracapacitors and other energy storage solutions. At one point, Lishen formed a JV with US-based EV manufacturer CODA and later with US-based LEV manufacturer ZAP.
Lithium Energy Japan (LEJ)—Japanese-based GS Yuasa formed a JV with Mitsubishi to develop lithium-ion batteries for automotive applications. Most recently through a partnership with Bosch, they have reinforced their focus on the automotive market.
Panasonic—Japanese-based Panasonic has been a major manufacturer of 18650 lithium-ion cells as well as NiMh cells. In 2008, Panasonic bought a majority stake in Sony Electronics company making the combined company one of the largest lithium-ion battery manufacturers in the world. Panasonic/Sony are one of the largest producer of 18650-type cells in the world.
Prime Earth Vehicle Energy (PEVE)—Formed in 1996 as a JV between Panasonic and Toyota Motor Company; Panasonic EV Energy developed NiMh cells for the Toyota Prius and many other HEV applications. In 2006, the company became Prime Earth Vehicle Energy and began offering lithium-ion cells for automotive applications.
Samsung—Korean Samsung SDI has been working on several chemistries including NMC since the late 1990s. In the mid-2000s, Samsung formed a JV with Bosch, called SB LiMotive with Samsung focused on the cells and Bosch on the packs. The JV only lasted a couple of years, but resulted in the battery design for the Fiat 500E battery. Today, Samsung SDI has several chemistries and prismatic cells available as well as complete pack solutions.

Saft—French lithium-ion manufacturer Saft has been working in lithium-ion batteries since the early 2000s. Saft offers large cylindrical cells in both primary (nonrechargeable) and secondary (rechargeable) products. Their market has focused on aerospace, satellite, military, stationary and grid, and automotive markets applications. As part of the JV with Johnson Controls, they were one of the recipients of the U.S. DOE funding projects. With operations in both Europe and the United States, Saft continues to find new applications for their cells.

Toshiba—Japanese-based Toshiba Corporation has been very active in the lithium-ion battery market, focused on their LTO chemistry-based Super Charged Ion Battery (SCiB™).

XALT Energy—XALT Energy began life as the Dow-Kokam JV, a JV between Kokam America, a distributor of NMC-based pouch cells from Kokam Korea, Dow Chemical Company based in Midland, Michigan, and French battery pack designer Dassault Systems. As one of the recipients of the U.S. DOE funding, they opened a lithium-ion manufacturing plant in Midland, Michigan. In 2012, Dow decided to exit the battery business and the company was reformed in 2013 as XALT Energy.

Design Guidelines and Best Practices

- Lead acid is a good fit for applications that requires standby type power, high power pulses, low cycle requirements, does not have weight or volume restrictions and low requires very costs.
- Nickel Metal Hydride (NiMh) has become the largest battery used in hybrid electric vehicles today. NiMh and nickel cadmium became popular due to their higher energy density as compared to lead acid. One of the challenges with nickel-based chemistries is the "memory effect" that the chemistries suffer from.
- Sodium-based chemistries offer about the same energy density as nickel-based chemistries, but are required to operate at very high temperatures in order to maintain their salts in a molten form.
- Lithium-ion has quickly grown to take over many of the battery applications that other chemistries have been used for in the past. Lithium-ion offers about twice the energy and power density of nickel-based chemistries, and about four times the energy density and power density of lead acid chemistries.
- There are many different form factors available for lithium-ion batteries. The three major form factors are cylindrical, prismatic can, and pouch-type cells.
- The most popular cathode chemistries today are lithium iron phosphate, lithium cobalt oxide, lithium manganese oxide, nickel manganese cobalt, and nickel cobalt aluminum.
- The major anode materials being used are graphite and carbons, both hard and soft carbon types.
- There are many cell manufacturers from which to choose, evaluate them based on your market needs, applications, and performance requirements.

CHAPTER 8

Battery Management System Controls

The Battery Management System (BMS), while it may have many other names, is the central control unit of the battery pack. It is essentially (and quite literally) the "brains" of the operation. The BMS is a combination of several component systems, including a host or master controller (a Printed Circuit Board (PCB)), a series of "slave" control boards (depending on system typography), sensors, and software that makes everything work together. Some people also expand the definition of a BMS to include the control electronics such as the switches, fuses, high-voltage front end, high-voltage interlock loop, and disconnects; however, for the purposes of this book, I have separated those components out into a separate chapter on System Control Electronics (Chapter 9).

What, then, does a BMS do in a lithium-ion battery system? In short, the BMS provides protection against overcharging, overdischarging, high temperatures, low temperatures, short circuiting, and other failure modes. In addition to protection, the BMS offers monitoring functionality; in fact the protection would be useless without the ability to monitor the state of the battery and cells. The BMS also communicates both internally within the pack as well as to the outside controllers and systems. Finally, the BMS provides optimization and maximization of the batteries performance, ensuring that the user can get the best performance out of the battery at any time. Overlaid on top of all of these are the software calculations that estimate the various factors within the battery such as the state of health (SOH), the state of charge (SOC), maximum voltage, etc.

That describes the "what" of the BMS; the "why" then is relatively simple. Why does a lithium-ion battery pack need a BMS? In short, the BMS ensures the safety and life of the battery pack. The BMS manages the amount of power and energy in the pack to achieve the desired lifetime of the pack. Most portable power batteries for laptops and similar devices only need to provide energy and power over a year or two and are generally operated in a narrow temperature range. How often do you leave your laptop outside in the middle of winter? On the other hand, the automotive and industrial batteries must last from 8 to 10 or even 15 years or longer and will need to survive in a very wide range of temperatures from the Arizona desert summers to the extremely cold winters of the northern regions.

Can a lithium-ion battery be used without a BMS? I have heard of some racing type EV applications where they claim to have no BMS in the system. But these are typically one- or two-use applications; the life of the battery may only need to last for one racing season and will be stressed to the extreme so warranty and longevity are not concerns. But overall, no production application will be designed without a BMS. Compared to other cell chemistries,

The Handbook of Lithium-Ion Battery Pack Design. http://dx.doi.org/10.1016/B978-0-12-801456-1.00008-7

lithium-ion needs to be managed in order to ensure its safety. A lead–acid battery, for instance, easily operates with no form of electronic controls as it is more easily able to accept abusive conditions without going into thermal runaway-type events. Lithium-ion, on the other hand, must be managed to ensure that the cells stay below their maximum and minimum voltage, temperature, and current limits.

This chapter will provide the reader with a very brief overview of the BMS, its functionality, and its hardware. For a more detailed review of the BMS and its operation, I would recommend Davide Andrea's well-written book *Battery Management Systems for Large Lithium-Ion Packs* (Andrea, 2010). Andrea covers all topics in great detail from general descriptions to how to design and build your own BMS.

BMS Typologies

There are two basic types of BMS system topologies, a centralized BMS and a distributed BMS. The main difference between the two is where the hardware is mounted. This is of course a major simplification but is relatively accurate. The host, or master, control unit usually encompasses the majority of the functionality of the ESS: (1) opening and closing the contactors, monitoring the temperature of the pack, and communicating with the cell control boards to monitor the cell temperatures; (2) monitors the voltage of the pack and cell control boards; (3) manages the heating and/or cooling (turning it on and off) based on the temperature readings; (4) manages the safety (opening and closing the contactors based on the voltage and temperature and SOx); (5) calculates, manages, and tracks the SOx functions, and communicates with the vehicle (or other master system). The controller is made up of two main components: a hardware controller board and a large amount of software "algorithms" that ensure the performance and safety of the complete system.

In a centralized BMS system, the main control board as well as the cell-monitoring control boards are all colocated in one unit with wiring harnesses spanning throughout the pack connecting to all of the cells (Figure 1). This minimizes the amount of hardware but increases the amount of wiring needed in the pack.

In the distributed BMS structure (Figure 2), there is a "host" or "master" controller that is centrally located and there are multiple separate boards to monitor the cells that are usually mounted directly to the cells or modules. This reduces the wiring needed as the slave boards tend to be connected in a daisy chain manner. But it also tends to drive up the cost as it increases the amount of PCB type hardware that is required. However, the distributed BMS design is often used to offer greater functionality and control within the system as each slave board controls only a limited number of cells.

There is some variability in the BMS architecture as the types of typologies actually vary across a continuum with the distributed at one end and the centralized at the other end.

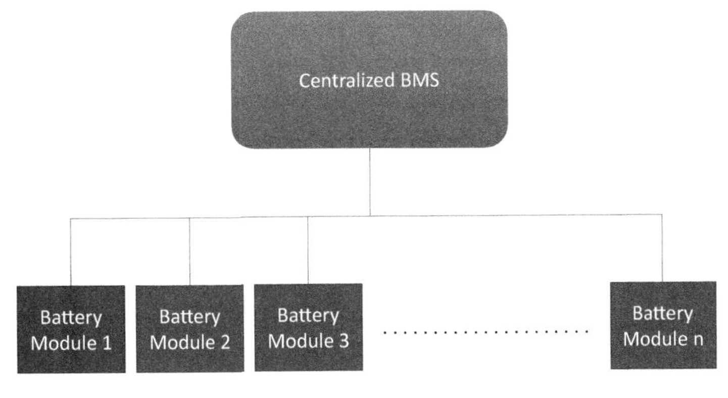

Figure 1
Centralized battery management system (BMS).

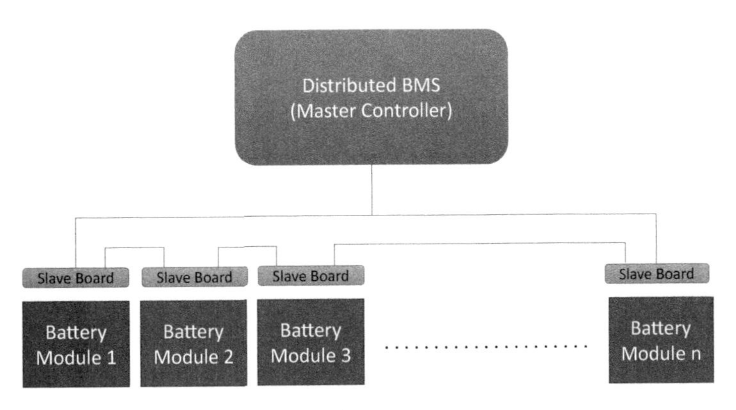

Figure 2
Distributed BMS.

There are several variations of these BMS system architecture and the final design will be determined by how the system is going to be used.

BMS Hardware

The BMS hardware board(s) are a critical part of the design process. It typically includes one or more PCB(s) that integrates all of the components that make up the controller board, including CAN, LIN, or other communications components, capacitors, resistors, current sensors (Figure 3), and most importantly the application-specific integrated circuit (ASIC) (Figure 4), all mounted onto a nonconductive substrate and connected with embedded conductive copper that is etched from copper sheets and then laminated onto the boards.

The slave boards may go by one of many different names, in fact depending on the designer. There are a wide variation in naming conventions, including voltage–temperature monitoring

(VTM) board, cell supervision circuit (CSC), and many others. But in any case, similar design focus should be made in the slave boards. Especially important in the slave board design is the balancing circuit and the waste-heat management of the boards. With a passive-type balancing system, these boards will convert energy into waste heat. The boards must be designed in such a way as to minimize the impact of the heat on either the cells or the board components itself.

Another important aspect of the electronic hardware design is taking into account the electromagnetic interference (EMI) and electromagnetic compatibility (EMC) of the components in the system. This is especially important when the control board is in close proximity with a charger, inverter, converter, or other high-voltage equipment. In these cases, it is important to provide adequate shielding of the hardware and the cabling that connects them.

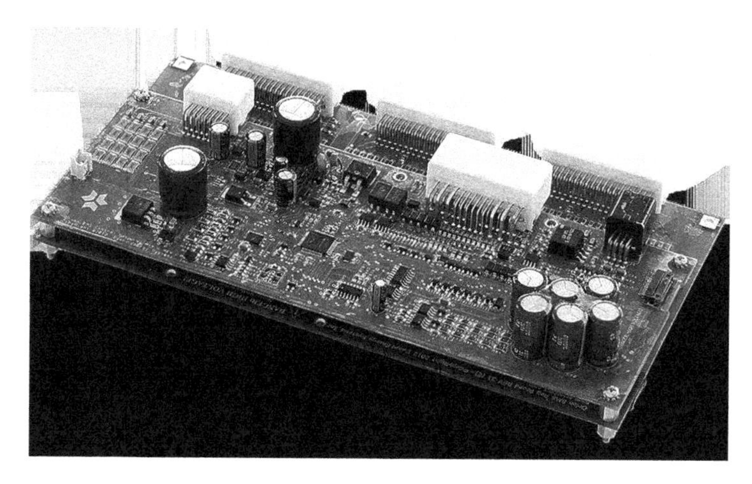

Figure 3
Printed circuit board (PCB) battery controller.

Figure 4
Texas Instruments application-specific integrated circuit (ASIC). *Courtesy:* http://www.kreiselelectric.com/en/technology/battery-system/.

Another option for the PCB slave board is the use of a flexible circuit board or flex circuit. This solution is essentially the same as the hard PCB board but instead of using a hard substrate, it uses a flexible plastic substrate, or a combination of the two. This is most often used in the CSC boards that mount directly to the cell modules.

Balancing

One of the other features of most BMS systems is their ability to maintain the cells in a pack at the same SOC, this is referred to as balancing the cells. The reason for balancing is that lithium-ion cells, just like any manufactured component, are manufactured to within a specification range but are not all exactly identical in their SOC or voltage when they are shipped from the factory. Additionally, lithium-ion, like most all batteries, suffers from "leakage" or "self-discharge" over time. In other words, if a cell is shipped at 3.7 V and 100% SOC, by the time it reaches the pack manufacturer it may be down to 99.5% SOC (purely for explanation purposes).

So for a large lithium-ion pack that is made up of hundreds or thousands of cells, the cells may all arrive at the pack integrator at very slightly different states of charge. While that may not seem to be a major concern because the variations are so small, it can create some major problems as the system begins operating. This is because the ability to charge and discharge is limited by the cell with the lowest (on discharge) cell SOC.

A simple example may help to clarify this. In the example below (Figure 5), we see three cells with two at about the same SOC and the third at slightly lower capacity. This is an unbalanced group of cells.

When the cells are fully discharged, cell #1 will be fully discharged before the other two and the pack will stop discharging as any further discharge will damage cell #1. This means that there will always be some remaining charge that is essentially unusable in the other cells as it will never be able to be fully discharged (Figure 6). Over time, this variation in the cell SOC will grow as the battery is cycled and eventually will lead to premature failure or end of life of the system. Additionally, the cell that is the weakest in this example will get more "use" than the others, resulting in it premature aging. In other words, this first cell will work harder

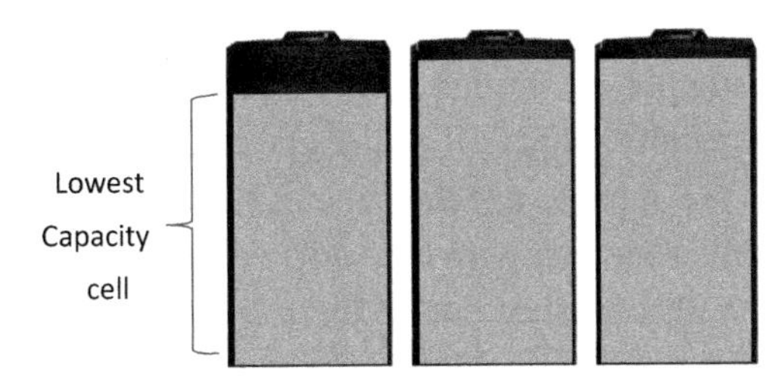

Figure 5
Imbalanced cells at beginning of discharge.

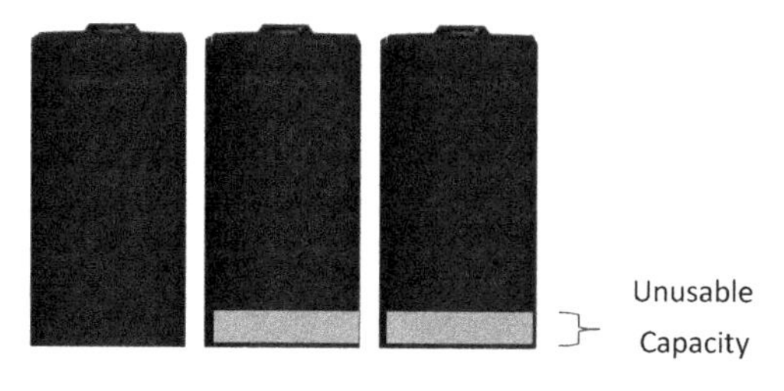

Figure 6
Imbalanced cells at end of discharge.

than the others, which will lead to an early death for both the cell as well as the complete battery system. So it is vital to ensure that the cells are as closely matched as possible.

However, this is exactly what BMS balancing is intended to resolve. Cell balancing is essentially the act of making all of the cells the same SOC. As Davide Andrea describes it, balancing is the term used for the process of bringing the SOC levels of cells in a battery closer to each other, in order to maximize the battery's capacity (Andrea, 2010, p. 23).

One additional thing to note here is that when cells are assembled into parallel configurations, they will automatically balance to each other. Each of the groups of parallel cells will still need to be balanced, but the cells within the parallel group will self-balance.

The other question that comes about when talking about balancing is when to do it. Most of the current BMS systems on the market today balance the cells while the ESS system is charging. This is done for a couple of reasons. First, balancing usually takes a significant amount of time and the battery must essentially be out of use when it occurs in order to accurately measure the capacity and voltage of the cells in the energy storage system.

Alternatively, for an HEV application, balancing could be done during a long freeway drive when the battery is essentially unused. However, this may generate some challenges while the BMS decides that you are on the freeway and can balance. This may be done by monitoring the speed of the vehicle and the duration that it remains at high speeds. Of course, this also assumes that you are on the road long enough to complete the balancing.

Alternative definition: another way to look at cell balancing is to imagine that you are at a party and a new game is being played. In this game, there is a table with three glasses on it. We need to fill and then eventually empty all of the glasses. But there are some rules we must follow. The first rule is that you have to use the special funnel device that allows all three glasses to be filled at the same time and with the same amount of water. The second rule is that you must stop filling the glasses once one of them is completely full, regardless of how full the others are. The third rule is that you must drain all the glasses at the same time using another funnel tool. And finally, the fourth rule is that you must stop emptying the glasses once one of them is completely empty. Ok, now that we have the rules. Let's take a look at what it might look like…(Figure 7).

Figure 7
Imbalanced cells.

In this game, we see that the water levels have been exaggerated greatly in order to facilitate the game. However, the result is clear. When you begin filling the glasses through the funnel (rule 1), the last glass will get filled compared to the top first (rule 2). Now you can see that the other two glasses still have plenty of room in them and in fact the first glass is not even close to being full (Figure 8).

Figure 8
Effect of imbalanced cells at full charge.

So using this example, we see that cell balancing can be done in one of two different manners. Either the water is removed from the fuller glasses or the water is transferred from the fuller to the less full glasses. The next section will describe these two actions, passive and active balancing.

Active versus Passive Balancing

There are two main methods for achieving cell balancing within a large lithium-ion battery pack, either active or passive balancing. The simplest explanation of the difference is based on what is done with the energy in the cell.

In passive balancing, the excess energy of the highest SOC cells is converted into heat energy and dissipated, essentially wasting the excess energy of the highest SOC cells. This can be done through several methods but the most common is to use a resistor to convert the energy into heat. Of course, this means that appropriate-sized resistors must be integrated into the slave boards that monitor the cells. This also means that you can use one resistor to balance multiple cells, although not at the exact same time.

The main benefit of passive balancing is that it is less expensive than an active balancing system, which makes it the preferred system for automotive applications. However, the detriment of this system is that energy is wasted as it is converted into heat energy. This creates a secondary problem as this heat generation must now be managed within the battery pack system. Another potential challenge to passive balancing is that it can be more time consuming than in an active balancing system (Figure 9).

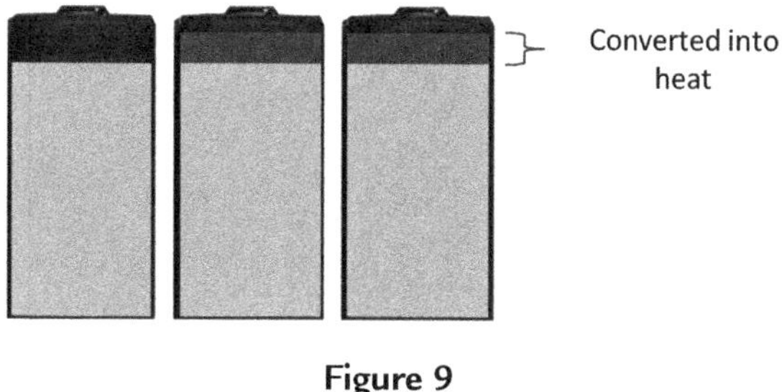

Figure 9
Passive cell balancing.

In an active balancing system (Figure 10), the excess energy of the higher SOC cells is moved to the lower SOC cells until all of the cells are at the same SOC. This may actually be done through a repetitive process whereby as more capacity is freed up, the charging will resume until the lowest cell again hits its maximum limit, then the balancing again resumes until all cells are at the same SOC.

The benefit of the active balancing system is that the excess energy is not wasted, instead moved into the other cells. However, the detriment of this is that the hardware required for

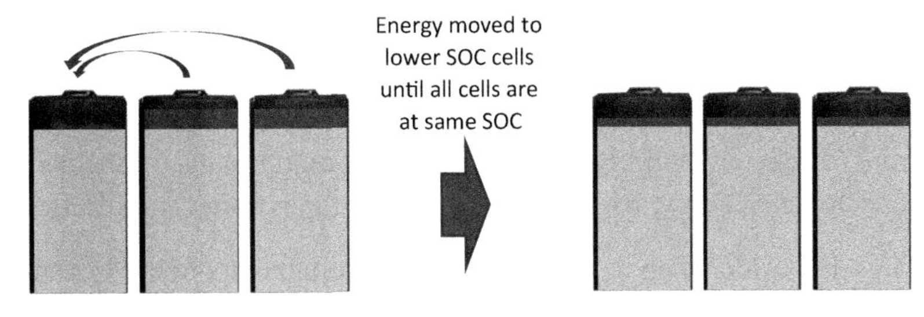

Figure 10
Active cell balancing.

active balancing is more expensive and can require more space within the pack. The electronics necessary to do this must also be attached to each cell, or integrated in the slave boards for a group of cells.

There have been a lot of evaluations of both systems over the past few years; however, current studies do not show the long-term benefits of using an active balancing system. In short with the current level of technology, there does not appear to be any major system benefits that can be achieved and those minimal benefits do not outweigh the added cost for an active balancing system.

Additional BMS Functionality

The other functionality of the BMS is no less important than balancing. In fact while it will significantly affect the life of a pack, the energy storage system could operate without any form of balancing. On the other hand, monitoring the temperature of both the cells and the pack, as well as the voltage of the cells and the pack is critical for maintaining the safety of the system.

The core job of the BMS is to ensure that the battery system does not allow the cells to operate outside of their safe operating range. This includes monitoring the pack current, cell and pack voltage, and the cell and pack temperatures.

Monitoring the pack current enables the system to determine how much power is instantaneously available for both discharge and charge (regeneration). Since driving the cells either over or under their maximum and minimum voltages can result in catastrophic failure, it is vital that the BMS has capability to monitor every cell in series within the pack (cells in parallel will be treated as a single cell in most BMS systems). These data can then direct the system as when to begin or stop a charge or discharge event. Finally, monitoring and managing the temperature of the cells and the overall pack is another vital piece of data as continued operation outside of these limits can not only reduce the life, but drive the cells into thermal runaway. The BMS is responsible for telling the system to send more cooling or heating to the cells.

The other important aspect of the BMS is its ability to communicate with the external system. Most advanced BMS systems will have the capability to both send and receive messages from the vehicle and or motor controller. Typically, the BMS will send requests to reduce or stop the battery current (discharge) entirely and will send data on the status of the pack itself such as remaining capacity and energy, which can be converted into range and life for the end user.

Finally, the BMS will decide when to open and close the contactors in the system thereby allowing current to flow from the pack to the electric motors or to flow from the charging system into the battery.

Software and Controls

The "unsung hero" of the BMS system is the software that controls it all. Most manufacturers will guard the core software very closely as this is the "core IP" of the entire BMS design. Most of the hardware is based on "off the shelf" components, but the software is custom designed and may consist of tens of thousands of lines of code.

The code may also be referred to as a series of algorithms. Essentially it is a series of mathematical formulas and calculations to understand all of the SOx states of the battery, how much energy and power is available for instantaneous use, what the current SOC is, how much SOC is left, and how much life is left. The algorithms are a very complex set of models that are usually based on a specific cell. Most often the BMS designers will operate the chosen cell in a controlled laboratory environment in order to understand how it operates under different conditions and then overlay this onto the code. Through a series of repetitive steps, it is possible for the software designers to end up with a set of algorithms that can accurately predict the performance of a cell under most conditions.

This is also the reason that it is generally not possible to take a BMS designed for a chemistry and integrate it into a system designed with a different chemistry. For instance, NMC cells operate at a nominal voltage of 3.7V, while LFP type cells operate at a nominal voltage of 3.3V and LTO cells operate at about 2.2V. So the algorithm must be designed such that it understands the maximum and minimum ranges that it can operate within. Now there are several BMS manufacturers out there who have developed multiple sets of software for their single set of hardware thereby making them usable with multiple chemistries.

Design Guidelines and Best Practices

- The BMS is the central control unit of the battery pack
- There are two major types of BMS typologies: Centralized and Distributed
- A centralized BMS has all of the hardware located in one spot in the battery pack
- A distributed BMS has a single master controller and then a series of slave boards attached to the cell modules
- Two major types of cell balancing are passive and active balancing
- Passive balancing uses resistors to convert excess energy into waste heat
- Active balancing transfers energy from a cell at a higher SOC to a cell at a lower SOC

CHAPTER 9

System Control Electronics

The same basic set of electronics hardware is used in both low- and high-voltage systems: main contactor, precharge contactor, high-voltage interlock loop (HVIL), manual service disconnect (MSD), fuses, bus bars, cell interconnect boards, and low- and high-voltage wiring harness. The assembly of these components is often referred to as the high-voltage front end (HVFE).

The type of electronics that make up the HVFE and are chosen will depend greatly on the total voltage of the system. A system with a voltage over 60 V is typically considered high voltage, which means that contact with voltage at or above this level will cause serious injury and potentially death to anyone who comes into contact with it. Therefore HV systems require an additional level of safety precautions, which must be taken in the design in order to ensure human safety while working with or around the system.

Systems below 60 V are considered low voltage; someone making contact with its components may get a shock, not a life-threatening injury. Systems of this type do not require the same amount of safety hardware and controls as are necessary in their HV cousins. Though it is a good practice to include appropriate safety systems even in an LV battery system.

While most automotive applications peak at around 400 V, there are some very-large MWh-size systems used in stationary energy and large industrial applications that can exceed 1000 V. The same concepts apply to these large systems; however, with these high voltages components must be sized appropriate to achieve the maximum voltage rating of the system.

As mentioned in the chapter on battery management systems (BMSs), another important aspect of battery design that must be evaluated during the design is ensuring that the system has proper electromagnetic compatibility (EMC) and electromagnetic interference (EMI) protections. This includes the use of proper shielded wiring where necessary as well as evaluating the design and location of any PCB and control boards in the system. The other item that can assist in managing EMC/EMI issues is the material that is selected for the enclosure. We will discuss this in more detail in Chapter 11 Mechanical Packaging and Material Selection. But the use of a metal enclosure can greatly help to mitigate the emissions of electromagnetic waves outside of the energy storage system.

The Handbook of Lithium-Ion Battery Pack Design. http://dx.doi.org/10.1016/B978-0-12-801456-1.00009-9

Contactors/Relays

A contactor is essentially an electromechanical switch, or relay, that will close in order to make the connection and to allow current to flow, or opens to stop the flow of current and voltage (Figure 1). This becomes another important design and safety component of the ESS as it will essentially start or stop the flow of current and voltage to and from the pack. The contactor is considered an electromechanical device as it is a mechanically driven device that makes the electrical connection.

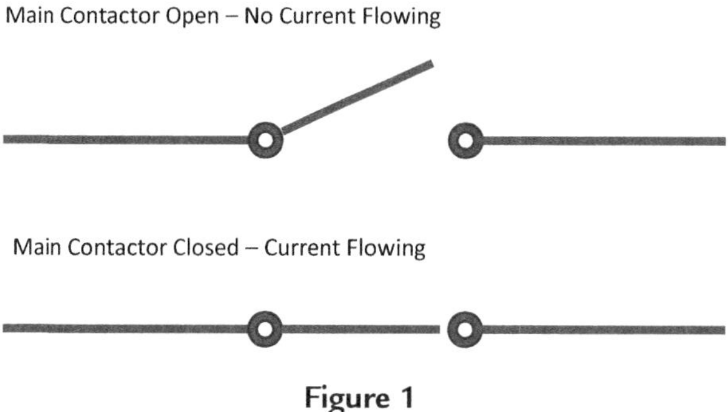

Figure 1
Contactor schematic in open and closed positions.

The contactor switch is often enclosed in a hermetically sealed unit filled within a nonconductive gas mixture in order to prevent sparking as the connection is made or broken. Sealing it in a nonconductive gas prevents sparking when the switch closes; additionally, as a sealed unit it can be used in corrosive environments such as the harsh automotive environment without becoming contaminated or corroded. Today there are several major companies that manufacture "off-the-shelf" type contactors for electrified systems. It is generally preferred to work with a manufacturer of these components in order to select the proper size for your system rather than trying to design one from scratch (Figure 2).

Figure 2
Tyco Kilovac 500 A 320VDC EV200 contactor.

Often used in conjunction with the contactor is a "precharge contactor" or "precharge resistor." This is a secondary switch that is connected in parallel with the main contactor. The purpose of this switch is to prevent a large inrush of current when the main contactor closes. It does this by closing the secondary contactor first and allowing a small amount of current to begin flowing into the system so that when the main contactor closes to complete the circuit it will do so without damaging any other circuitry or causing the main contactors to become welded together in a closed position, which could be very dangerous (Figure 3).

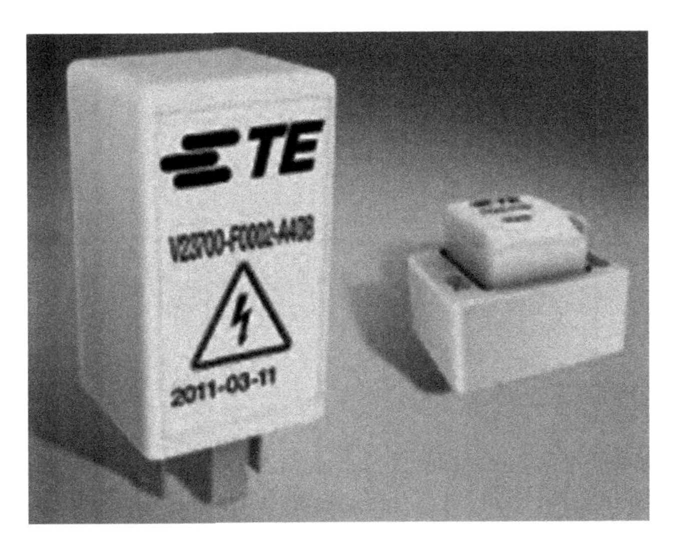

Figure 3
TE Connectivity precharge contactor.

In Figure 4 below, three different schematics for the main and precharge contactors are shown. In the top figure, both the main and precharge contactors are open so no current is flowing through the system. This is basically in the "off" position. In the next schematic the bottom, precharge contactor is closed and thereby allowing a small amount of current to begin flowing into the system, while the main contactor remains open for a predetermined period of time. Finally, in the bottom schematic, the main contactor has also closed and now the system is allowing full current to flow and the system is in the "on" position allowing current and power to be added or removed from the energy storage system.

In most systems, there is only one contactor and precharge contactor. However, in some of the very large systems that are made up of several interconnected "strings" or packs in parallel, each string may have its own set of contactors thereby allowing for each one to be run independently.

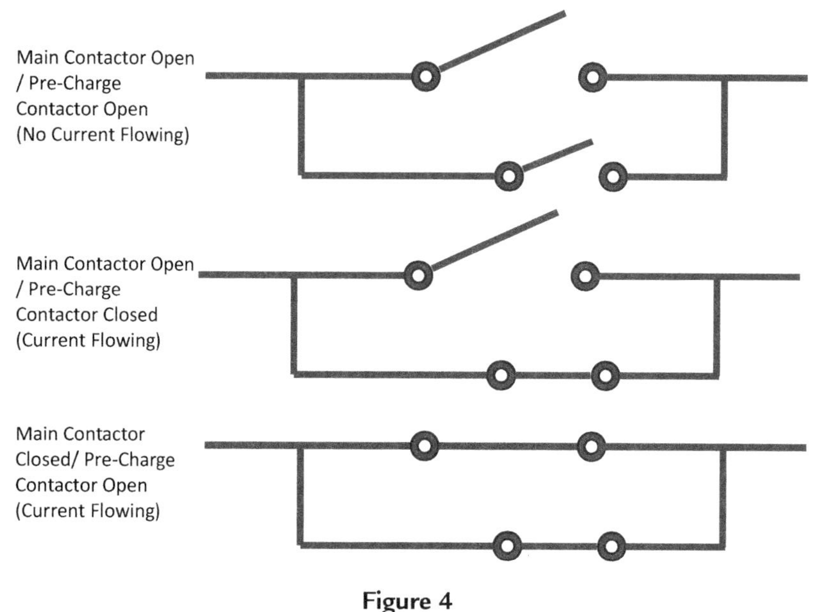

Figure 4
Main and precharge contactor positions.

High-Voltage Interlock Loop

Many high-voltage energy storage systems also include a safety feature that is called the HVIL. The HVIL essentially creates a closed circuit when the battery pack is sealed; if one part of the pack is "opened," the circuit is broken and the contactors are opened in order to prevent current from flowing. But it is important to note that the HVIL is not a single component that can be installed in the battery system, but rather a series of components, software, and controls along with an integrated pack design that will not allow for the pack to be opened until the HVIL is disengaged. The purpose of this is to ensure that the energy storage system is safe to work on when it is opened up for service or maintenance. One method of accomplishing this is through connecting the HV cables to an interlock circuit that reports the status to the battery pack controller. With this type of system, when the BMS controller recognizes when one of the interlock circuits is opened, the battery pack controller will send a message to the contractors to force them to open therefore the high-voltage bus can be discharged and the pack opened safely. The other method of doing this is to connect the HV power through an MSD device that has a fuse integrated into it. In this case, once the MSD is removed the circuit is opened and voltage will cease to flow.

Fuses

Most automotive high-voltage energy storage systems are also designed with an MSD as an integral part of the HVIL circuit (as mentioned above). The MSD, also called a mid-pack service disconnect or just a service disconnect, is a fuse that is usually placed at or near the "middle" of the pack so that when the fuse is removed the potential or energy of the pack will also be cut in half and the HVIL circuit opened, so that the contactors will open. As mentioned above, the MSD is typically designed as part of the HVIL so that the HV power cables are connected through the MSD (Figures 5 and 6).

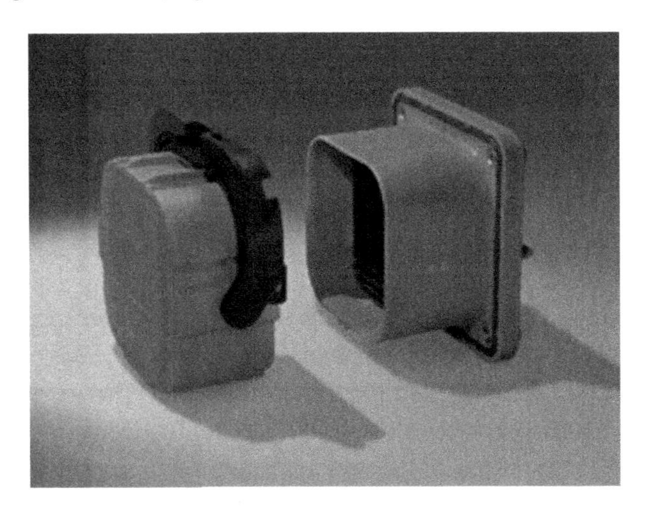

Figure 5
Manual service disconnect. *Image used courtesy of TE Connectivity.*

Figure 6
Manual service disconnect. *Image used courtesy of Delphi Electronics.*

Depending on the size of the pack, there may be a need to include other fuses into the overall design along with the MSD. In low-voltage packs, it is generally not necessary to include an MSD, but instead there may be a need to include one or more fuse depending on the system design.

Additionally, when it comes to fusing, some OEMs have integrated small fuses directly into the control boards that interconnect the cells. This provides the benefit of additional safety in the event that any single cell, or set of parallel cells, experiences a failure and voltage spike.

Large energy storage systems may not include an MSD-type fuse, but will generally include a fuse of some sort into the pack design as a safety feature. For the larger systems, the fuse may actually be integrated into a battery disconnect unit (BDU).

Battery Disconnect Unit

Most of these HV electronics components and subsystems are often packaged together into one location in the battery pack. In fact, some component suppliers and OEMs package them into a single physical unit that may be referred to as the BDU. While the BMS is typically mounted separately from the BDU, the BDU serves as the interface to the electric motor, vehicle controllers, and any communications outside of the vehicle. Some component suppliers have designed these into "standard" off-the-shelf type units that can be purchased as a single unit and installed in your energy storage system. In these off-the-shelf units, the main contactors, precharge contactors, fuses, HVIL circuitry, and other electronics hardware are all integrated into a single unit.

Several other companies such as Delphi and Tyco Electronics provide "off-the-shelf" HV units. Depending on your team's HV electrical experience and the requirements of the application, this could be a good fit and could save your team's significant engineering time and effort (Figure 7).

Figure 7
Off-the-shelf high voltage (HV) electronics by Delphi Electronics.

Connectors

The other thing to make sure if evaluated when designing the controls and electronics systems are the connectors. While this may seem like a "no brainer," the lithium-ion battery industry is only just beginning to get to some level of standardization so there are still many solutions available and each has different costs/benefits—and they are not all compatible with each other! Additionally there are many portable electronics application connectors that may be evaluated but many do not yet meet the rigid performance or life requirements of the automotive application (Figure 8).

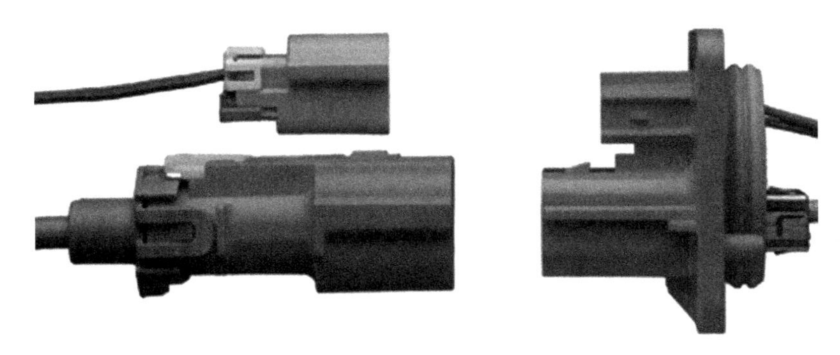

Figure 8
Automotive EV battery connectors. *Image used courtesy of Delphi Electronics.*

Connectors must be evaluated for the level of sealing that is required in the pack, their ability to withstand abusive environmental conditions, the amount of EMI shielding, and of course their ability to transmit the proper currents, voltages, and communications without adding excess costs to the systems (Figure 9).

Figure 9
TE Connectivity AMP + high-voltage connector. *Image used courtesy of TE Connectivity.*

Another connector issue that should be properly evaluated is the module-to-module and cell-to-cell connections. Some connector manufacturers have developed some interesting connector options that can improve ease of installation and serviceability. One such example is Amphenol who have developed a unique "Radsok" type connector that is designed for easy installation and disassembly. One of the current challenges of them is that they are still a bit more expensive than more traditional connector options.

Two final thoughts on wiring and connectors: It is especially important to ensure that the connectors that are chosen have a positive locking mechanism built into them. This is very important in applications that are transportation based but should be considered in all battery systems. By integrating interlocking connectors you are ensuring that the connections do not disengage or separate over time. You may also notice that much of the "standard" high-voltage wiring and cables has a specific orange coloring; this is because SAE standard J1673 defines the requirement for identifying these in automotive applications.

Charging

While we will not go into great detail on battery charging in this book as it is considered to be on the "other" side of the battery system, but we should still spend a short time discussing the impact of different charging solutions on the battery. Battery charging is typically referred to as level 1, level 2, or level 3 charging (Table 1). In essence, this refers to the voltage being used to charge the battery. The EV charger is also sometimes referred to as the electric vehicle supply equipment (EVSE).

Table 1: Charging levels

	Voltage (V)	Current (A)	Power (kW)	Type
Level 1	110	16	1.9	AC
Level 2	208/240	32	19	AC
Level 3	480	400	240	DC

The impact on the battery of each of these levels is slightly different. For instance, level 1 charging is the easiest for the battery to use at it has very low c-rate charging. It is also about the standard voltage level in most US households, which means that you can charge your EV from a standard household outlet. Most EV manufacturers include level 1 chargers with the vehicles and have integrated the standard SAE level 2 connectors.

Level 2 charging is the most commonly used as it is about the same voltage as most household and industrial businesses. Level 2 charging is relatively safe on the battery as it will still take between 4 to 6 h to charge using this method. SAE has also developed a standard for level 2 charger connectors as shown in Figure 10 below.

Figure 10
SAE J1772 Level 2 connector.

Frequent level 3 fast charging may cause the dissolution of the cathode due to the off-gassing of electrolyte. This may cause reduced life of the battery system. Many vehicles that are capable of level 3 charging will limit the number of times that a vehicle can use a level 3 charge before requiring a level 2 charge. The reason for this is to allow for balancing and to allow the battery time to cool down. Some battery chemistries have less impact from fast charging than do others. There are basically two types of charger connectors associated with level 3 charging. In the United States, the SAE has developed a combined level 2–level 3 connector as shown in Figure 11 below. In Japan, a different connector standard has emerged and is known as the CHADEMO connector. The CHADEMO connector was developed and standardized by a group of Japanese suppliers and manufacturers including The Tokyo Electric Power Company, Nissan, Mitsubishi, and Fuji Heavy Industries with Toyota later joining.

Changes and improvements in electrolytes may help to improve battery fast charging performance. In fact, several battery manufacturers have already claimed to have made improvements in their cell chemistries in order to enable more frequent fast charging.

Another electronics technology that is getting more interest is inductive charging. Again, while this may not specifically require a major change to the battery design, it will require the integration of a secondary coil in the battery pack in order to complete the circuit and create the electromagnetic field that enables the wireless charging. An inductive charging system is a wireless means of charging a battery by creating an electromagnetic field between two coils.

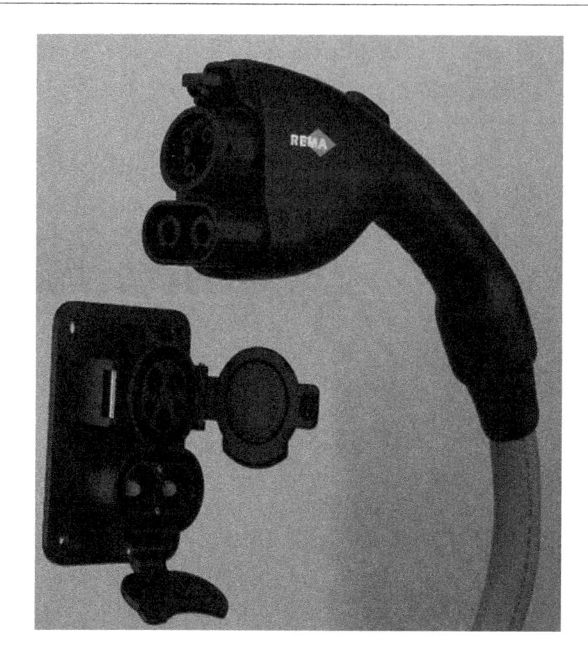

Figure 11
SAE Level 3 combined connector.

Today these systems still have relatively low efficiency, meaning that a good portion of the energy that is trying to get transmitted is getting "lost" in the process. However, major improvements are being made in this technology, which may make it mass-market-ready in the very near future.

Design Guidelines and Best Practices

- The second half of the control electronics system is the high- and low-voltage electronics. This includes the contactors, HVIL, fuses, MSD, and the high- and low-voltage wiring harnesses.
- Contactors and precharge contactors or relays are electromechanical switches, or relays, that will close in order to make the connection and to allow current to flow, or open to stop the flow of current and voltage.
- The HVIL is a system of components and systems that creates a closed circuit when the battery pack is sealed; if one part of the pack is "opened," the circuit is broken and the contactors are opened in order to prevent current from flowing.
- MSDs are included in most automotive high-voltage batteries. The MSD is a fuse that is usually placed at or near the "middle" of the pack.
- The battery disconnect unit is the combination of the BMS, contactors and relays, fuses, and electronics components integrated into a single unit. Several suppliers offer "off-the-shelf" HV electronics power distribution units.

Design Guidelines and Best Practices—cont'd

- Connectors come in many shapes and sizes, but connector selection is an important part of a system design. Connectors should be evaluated to understand if they need to be sealed, shielded from EMI/EMC, or if a standard connector exists.
- Electric vehicle charge equipment (EVSE) ranges from level 1 charging at 110 V, to level 2 charging at 240 V and level 3 charging at about 480 V.
- Standard charger connectors are available from many suppliers in both SAE and CHADEMO types.

CHAPTER 10

Thermal Management

In this chapter, we will discuss the different manners in which a battery pack can be cooled and, if necessary, heated to maintain the cells at optimum operating temperature. When it comes to lithium-ion cell temperatures, the most important thing to remember is that lithium-ion cells like to be maintained at about the same temperature range that people are comfortable, which is about 23 °C (73 °F).

Thermal management of a battery is, in its most simple form, the integration of some form of heat exchange device into the battery system in order to maintain the battery cells at a near constant temperature. There are many potential methods that can be used to achieve this, but it typically involves moving heat through some medium away from the cells and out of the pack. The complexity of the thermal management system will be highly dependent on three factors. The first is the duty cycle under which the pack will operate. If it is intended to be used as a high-power application, the battery cells will generate more heat than if it is used as an energy application. The second factor is the region where it will be used in. If the pack is intended to be used in a region of the world that frequently has very high ambient temperature, it means that the batteries will begin their cycle at an already elevated temperature level. Finally, the third factor is the cell itself. Different chemistries perform differently under high load and temperature operation. Therefore, it is important to consider all of these factors when beginning to design your battery's thermal management system.

The thermal management system should be able to maintain a temperature difference of about 2–3 °C from the coolest cell to the warmest cell. At the worst-case condition, usually for larger packs the difference can be as much as 6–8 °C. The reason why this is important is that a large temperature gradient between the cells will cause the cells to age at different rates. So the hotter cells will age faster than the cooler cells, and if there is a large gradient, this could mean that the battery's calendar life will be reduced prematurely.

There are three different types of heat transfer that need to be considered in battery design: conduction, convection, and radiation. Conduction refers to a direct transfer of heat energy from two objects that are in direct contact. Convection occurs when heat is conducted through a liquid medium to a heat-sinking device. Radiative heat transfer refers to heat energy that generated through electromagnetically thermally charged particles of matter that radiate from one source to another, generally through the air. All three methods of heat transfer must be considered in the battery system design, but conduction and convection will have the greatest impact on the thermal system design.

The Handbook of Lithium-Ion Battery Pack Design. http://dx.doi.org/10.1016/B978-0-12-801456-1.00010-5

For example, as the cells are discharged they generate heat, that heat will be transferred via conduction to the bus bars and any other components that are in direct contact with the cells. Convective heating and cooling is most typically seen through the use of a liquid cooling plate; however, the movement of cooled air can also provide convective cooling. This is where most of the cooling will take place in the battery system. Finally, radiant heating must be considered in a couple of important areas. First, the radiated heat from the cell to other components that are not in direct contact with the cells can impact adjacent cells. Second, the heat that is being generated from components such as the electronics will impact the cells if it is not properly managed.

Sources of heat generation inside a battery system come from the chemical reaction within the cell (the main area of heat generation) as the lithium-ions move back and forth during operation; balancing of cells (usually while parked); electronics within the battery pack; and thermal management system. While the lithium-ion cells are the primary source of heat generation inside a pack, the influence of the electronics must not be disregarded as it can be a significant source of heat creation that must be accounted for in your thermal management system design. If adequate shielding and consideration of the placement of these electronics are not considered, then the heat generated by the electronics may actually have a negative impact on the cells life.

Figure 1 below offers some simple examples of each of these three methods of heat transfer. In this example, a pouch-type lithium-ion cell is shown with a cooling plate on the right and

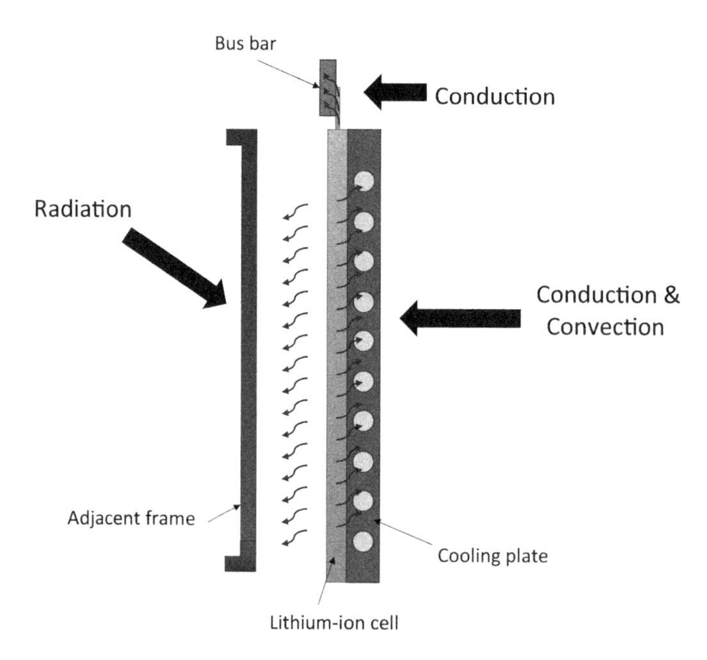

Figure 1
Sources of heat in a lithium-ion battery.

the adjacent mounting frame on the left. The heat transfer from the cell to the cooling plate uses conduction both in the direct conduction of heat from the cell to the cooling plate and to the liquid cooling channels. These channels then use convection to remove the heat through a liquid medium away from the cells. At the top of the cell, a simple copper bus bar is shown being connected to the tab of the cell. This demonstrates the heat transfer through direct conduction from the cell to the bus bar. Finally, the frame piece to the left of the cell may be used for mechanical structure of the module; however, even if it is not directly in contact with the cell it will experience radiative heating from the cell.

Lithium-ion chemistries tend to operate best between about 10 and 35 °C; this is referred to as the optimal temperature range. This is where you want the batteries to be at most of the time. However, most all lithium-ion chemistries will still operate down to about −20 °C and up to about 45 °C; this is known as the operational range. In this temperature range, no reduction in battery life would be expected to be experienced during normal operation. Between −20 and −40 °C the electrolytes may begin to freeze and the cold temperatures increase the impedance within the cell thereby resisting the flow of ions and reducing capacity and performance, and above 60 °C many lithium-ion cell chemistries begin to get more unstable; this is known as the survival temperature range (Figure 2).

This brings us to another concept that we should consider, at least briefly, in our thermal system design—thermal mass. Thermal mass is essentially the amount of heat that each object within the pack that has mass (all of them!) can absorb. Much in the same way that a battery can only store so much energy, all of the components in the battery pack can absorb a certain amount of heat, which is directly relative to their mass and the specific thermal properties of that particular material. For example, a copper bus bar will be able to absorb more heat than a plastic cover. Above its maximum ability to absorb heat, the material will begin to breakdown, plastics will melt and/or burn, and metals will begin to melt. Thermal mass will not generally drive major design considerations, other than in some of the components that are directly in contact with the areas of heat generation. However, in a passively

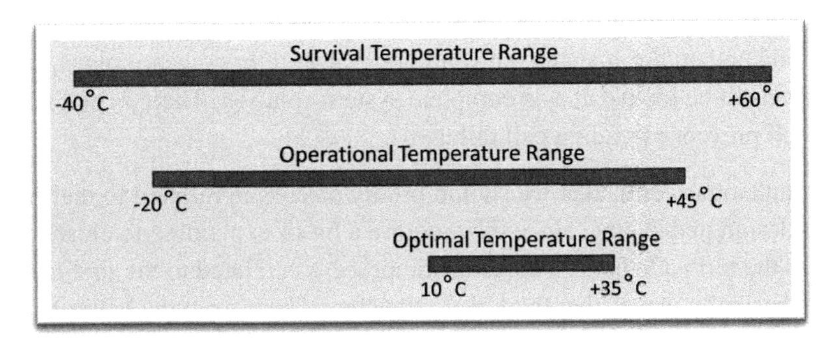

Figure 2
Lithium-ion cell temperature ranges.

cooled pack or an uncooled pack, this concept must be examined to evaluate its impact. Heat capacity may also come into effect if the application is being used in regions of high ambient temperatures. This is because in these regions the battery pack may absorb some of this ambient heat, which will cause the thermal system to have to work much harder in order to remove this heat. An example of this may be seen in regions such as the state of Arizona in the United States. Here, with summer ambient temperatures that may be in range of 38–44 °C (100–110 °F) or higher, the battery pack will absorb that heat and therefore will have to work to cool down the pack from these initial high temperatures back down to 23 °C in order to preserve the life and performance of the battery system.

At temperatures above about 90 °C, a polymer-based separator may begin to melt and breakdown, and between 90 and 130 °C the separator will continue to breakdown until a series of internal short circuits between the anode and the cathode are experienced; at this point, the cell will begin to move toward what is known as "thermal runaway." In effect, thermal runaway means that the cell becomes hot enough to create self-sustaining heat generation and failure is imminent in the form of cells venting and/or explosion (often referred to as "rapid disassembly"). There is no way to stop a cell in thermal runaway once the threshold has been surpassed. The exact temperature that a cell reaches the thermal runaway threshold is different for different chemistries. Some may reach it as low as 120 °C, while others may be able to exceed 140 °C before reaching this event.

The other factor that occurs during a thermal runaway is that as the chemistry inside the cells begins to breakdown, side reactions begin occurring rapidly generating oxygen. This oxygen generation may continue to fuel the burning of the cell. In this instance, a sealed pack design may offer some benefit due to the fact that no new oxygen is allowed to enter the pack. Therefore, once the oxygen inside the pack is consumed, there may not be enough oxygen to sustain a flame. One final comment on thermal runaway is that the temperatures that can be experienced during these events may be able to exceed 600–800 °C depending on the size of the cell and how many cell fail. There are few solutions that can prevent a cascading failure in the event of a thermal runaway with current chemistries. However, system design should take these safety factors in account as much as possible and attempt to either manage to thermally isolate the failing cell and/or manage the venting of the cell. In any case managing cascading cell failures needs to be looked at as a complete system solution. There is no "silver" bullet that can by itself prevent cascading cell failures.

There are a couple other terms that we should briefly discuss in relation to thermal management systems design and testing, since they require a bit of explaining to clearly understand. The first one is the term adiabatic. The adiabatic process is related to the first law of thermodynamics that basically means that there is no transfer of heat from the battery to the outside environment. In relation to lithium-ion batteries, this may be used in relation to thermal characterization testing on a lithium-ion cell. For example, when the cell is being tested, it is done in a thermal chamber that has the capability to operate at varying temperatures. So when the cell is being tested, the thermal chamber is maintained at the exact same temperature as the cell throughout the test. This allows for the accurate definition of how much heat is generated by a cell during a specific operational cycle.

The other two terms that I want to explain are exothermic reactions and endothermic reactions. An exothermic reaction is one that occurs when energy is discharged and heat is released into the environment thereby causing heating. An endothermic reaction occurs when energy is absorbed and heat is absorbed from the environment thereby causing cooling. In lithium-ion cells, exothermic reactions occur when the cell is being discharged (releasing energy) thus generating heat that must be managed through the thermal management system of the battery.

Why Cooling?

As mentioned above, in order to maximize the potential of lithium-ion cells, they need to be maintained at about 23–25 °C (73–77 °F) throughout the majority of their use cycle. However, under operation, the cells experience an exothermic reaction—they begin generating heat due to the rate at which the chemical reaction occurs within the cell and the related increase in cell resistance. This reaction in combination with high ambient (outside) temperatures means that your battery design must be able to cool the batteries down and maintain them within their optimal operating range in order to ensure the performance and life of the overall battery system.

Additionally, high discharge rates generate the exothermic temperature increases within the cells. And when these discharges come frequently, it means that the cells do not have time to cool down between these pulses, which again drive higher temperatures. In addition to this, there is less time for the thermal management system to engage and reduce the temperature of the cells back down. Think of this as a stair-step effect: when you hit your accelerator, a rapid discharge of the battery occurs; with frequent stopping and accelerating events the battery thermal management system will not have time to cool down the battery from the last discharge, causing the battery to gradually but steadily increase in temperature. An example of this is the traditional HEV, which will continually discharge and then regeneratively charge the battery regularly during a driving cycle. In this type of usage cycle, the battery may not have time to cool down the cells before another discharge–charge cycle is initiated. This will cause a slow and steady increase in the temperatures within the pack. Allowing the pack to sit unused for a period of time will generally allow the system to reduce the temperature back down to its normal operating range. Figure 3 below offers an example of what this type of temperature increase may look like. While this is not actual performance data, it is consistent with the type of heat generation that could be expected to be seen in this type of application.

As mentioned earlier, in high-temperature environments where the ambient air temperature is already in the 30–35 °C range or higher, it becomes even more important to be able to reduce

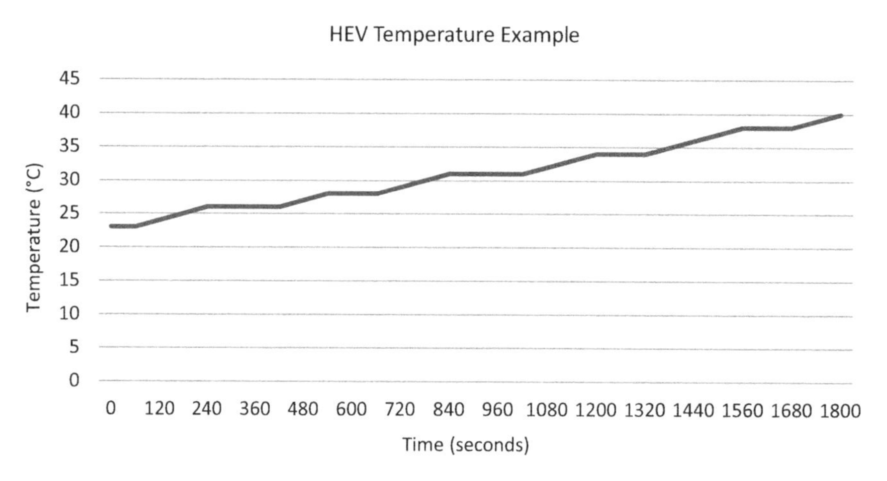

Figure 3
HEV temperature example.

the temperature down below the ambient in order to ensure life, performance, and against catastrophic failure. Constant operation at high temperatures reduces cell life and speeds the aging process. This was demonstrated in 2012 when a group of Nissan Leaf owners in the Arizona region complained to Nissan about the low electric range and reduced battery capacity that they were getting with their vehicles after only a short period of time (Gordon-Bloomfield, 2012; King, 2013). While there were not initially many of these complaints, there were enough to generate much unwanted national media attention for both Nissan and the industry in general. In the end, Nissan responded to these criticisms in an open letter on their MyNissanLEAF.com forum stating that:

> *Battery capacity loss of the levels reported may be considered normal depending on the method and frequency of charging, the operating environment, the amount of electricity consumed during daily usage and a vehicle's mileage and age.*
>
> ***(Carter, 2012; Bailo, 2012)***

They then went on to state that:

> *Battery data collected from Nissan LEAFs to date currently indicates that less than 0.3 percent of Nissan LEAFs in the U.S. (including vehicles in service dating back to December, 2010) have experienced a loss of any battery capacity bars. Overall, this universe of vehicles represents a very small fraction of the more than 13,000 Nissan LEAFs on U.S. roads. Also, data received globally from other LEAF vehicles shows that this condition typically occurs to high-mileage cars or those in unique operating situations.*
>
> ***(Bailo, 2012; Carter, 2012)***

In essence, Nissan's evaluation of the vehicles and the way they were being used found that the users who were experiencing this reduction in capacity tended to be driving about 150% of the average US annual mileage, some of whom were also using frequent fast charging.

CAUTION

To prevent damage to the Li-ion battery:

- **Do not expose a vehicle to ambient temperatures above 120°F (49°C) for over 24 hours.**
- **Do not store a vehicle in temperatures below −13°F (−25°C) for over seven days.**
- **Do not leave your vehicle for over 14 days where the Li-ion battery available charge gauge reaches a zero or near zero (state of charge).**
- **Do not use the Li-ion battery for any other purpose.**

Figure 4
2012 Nissan LEAF Owner's Manual battery warning (page EV-2).
© 2012-2013 Nissan North America, Inc. Nissan, Nissan model names and the Nissan logo are registered trademarks of Nissan.

So even while the batteries were performing to their design specifications, the continuous high temperatures and more extreme usage drove a reduction in battery capacity and therefore reduced range. And even though Nissan's owner's manual (Nissan, 2011) tried to clearly identify the major areas that will reduce battery life, they suffered much undeserved negative attention (Figure 4).

In the end, Nissan ended up changing their warranty between the 2012 and the 2013 model-year vehicles in order to account for these types of occurrences. Prior to the 2013 model-year vehicles, the warranty specifically excluded gradual loss of charge/capacity. To their original 8-year/100,000-mile warranty that covers defects and flaws, they added a 5-year/60,000-mile warranty that covers capacity. If the Leaf loses more than 30% of its original capacity in the first 5 years or 60,000 miles, Nissan will repair or replace it.

In addition, Nissan began working on the release of a new battery pack design that was intended to resolve these issues. To resolve these issues, Nissan began looking at alternative battery chemistries that would perform better in the high temperature regions of the world (Gordon-Bloomfield, 2013). Additionally, in mid-2014, Nissan announced that they would be offering replacement battery packs for earlier models at a price of $5499 USD (O'Dell, 2014). This is most interesting as this calculates out to a price of about $229 per kWh ($229/kWh), which is much below the current market pricing. Nissan is very likely taking a loss on these replacement packs in order to help inspire current and future customers, but it will be at a cost.

To tie this back in with the thermal management system of the battery, the reader should recognize that the Nissan Leaf battery pack (as will be discussed in more detail in a following chapter) actually has no active thermal management system at all. Instead, the Leaf battery pack designed by AESC uses the metal enclosures of the module and the pack to act as a heat sink to spread the heat out. But in hot climates or in frequent rapid discharges, there is no means to increase the rate of cooling in order to more rapidly respond to the increasing temperatures. While this provided a low cost and fast to market solution, as we have seen it also caused some other problems on which they are still working hard to resolve.

Finally, we should talk about the high-voltage electronics and balancing systems when we talk about thermal management. In many systems, the high-voltage electronics can be a major portion of the heat generation of the pack, especially in smaller packs. Additionally, cell balancing by its very act is the transference of electrical energy into heat energy. In essence, as you balance the pack you also heat it up. In the cases where level 3 charging is used, most battery systems are designed to turn the thermal management system on during these charging events due to the rapid heat generation that will otherwise ensue. Without the cooling system being engaged during a fast-charge event, the battery may generate enough heat to reduce the amount of energy it will be able to hold and could generate enough heat to actually cause the battery system to shut down until it cools. Think of this as trying to use a fire-hose-size nozzle to fill your gas tank; the tank will simply not be able to take that volume of fuel at once. In the battery system, it is the not the speed but in fact the heat that is being generated that causes the problems and stresses the lithium-ion cells.

In the case of the electronics within the battery system and their heat generation, it is generally easy enough to run cooled air across the electronics to help cool them down. However, the designer in these instances needs to ensure that the air is flowing to the lithium-ion cells either *before* or *in parallel* with the flow to the electronics. Most electronics have a much wider operating temperature range than the batteries do, so it is more efficient to cool the cells first and the electronics second. In liquid-cooled battery packs, it is somewhat more difficult to ensure the cooling of the electronics unless cooling plates are designed into the pack. But in both situations the location and placement of the electronics require special attention by the energy storage system designer.

All of these need either some amount of thermal management or at least to have a system design that does not negatively impact the performance of the batteries.

Why Heating?

On the other end of the spectrum is heating the battery. As a general rule, battery system designers do not want to heat a battery except for at very low ambient temperatures and even then it should not be a "rapid" heating.

Most liquid electrolytes used in modern lithium-ion batteries will begin to experience reduced power at about −10 °C and will begin to freeze at between −20 and −30 °C, which makes them unable to provide power at very low temperatures.

In a liquid-cooled system, a heat pump can be added to the overall system to provide warmed liquid through the cooling loop, which will slowly heat up the batteries. Other methods may also be employed such as using a thin-film heater. In this instance, the heater may not actually be used to heat, but rather to slow the rate of cooling such that if the vehicle sits for a weekend the cells will only be down to 25 °C after several days. This means that the system would have full performance, however, would suffer from some energy loss as the thin-film heater would need to be powered by the battery itself. However, this would be only a minor and temporary capacity loss.

Active Thermal Management Systems

Active thermal management involves using some medium such as air, liquid, or refrigerant that is forced through the pack and over the cells to reduce temperatures. The two most common methods are air cooling using chilled air that is directed throughout the pack and over the cells and electronics to reduce the temperature of the cells and liquid cooling. This generally requires the integration of a fan, ducting, and heat transfer plates of some sort. The benefits are that it can be relatively effective in responding to rapid changes in temperature and has a lower weight than a liquid-cooled system. The other benefit is that the cooled air is directly flowing across the cells. The disadvantages are that air is not as effective a cooling medium as liquid. And depending on air-flow design, it can cause the cells at the beginning of the air flow to be cooler than the cells at the end of the air flow because as the air passes by the initial cells, it begins to pick up the heat so by the time the air passes by the last cells it is warmer than when it in initially entered the pack. This uneven cooling can cause the cells to age at different rates, thereby reducing the life of the pack (Figure 5).

In addition to using chilled air, some systems will pull ambient air from the outside of the pack, or even inside the cabin, and circulate that air through the pack. In this case, the air

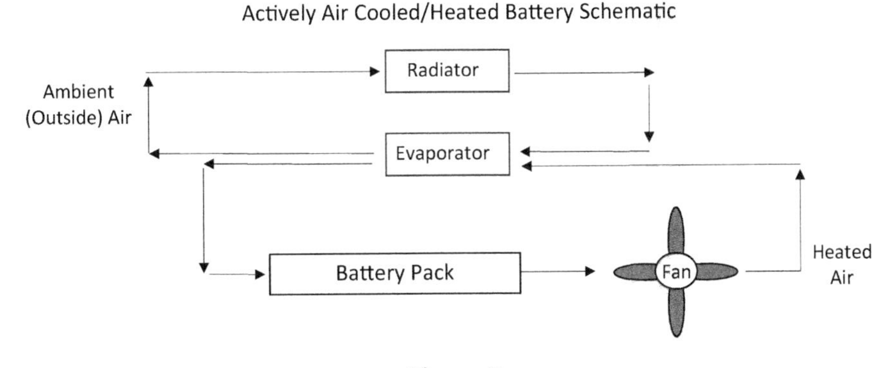

Figure 5
Active air cooling schematic.

Passively Air Cooled/Heated Battery Schematic

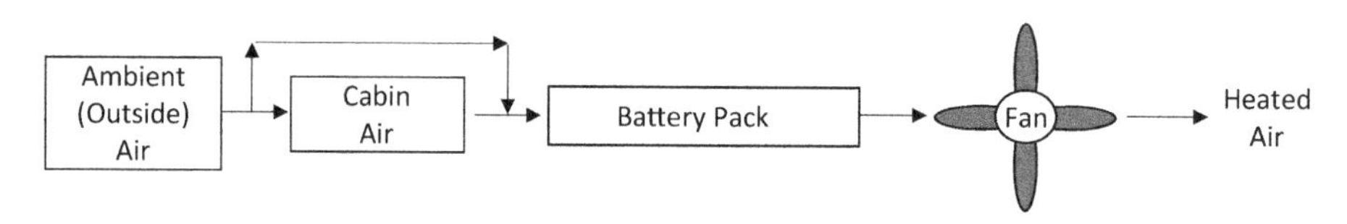

Figure 6
Passive air cooling schematic.

must first be filtered and the cooling ability is limited to reducing the temperature only to what the ambient temperature is. So in this instance, if the ambient temperature is at 30 °C the thermal management system will only be able to cool down the battery to 30 °C (Figure 6).

Another challenge of air cooling is that it makes the system design more difficult in the event that the pack needs to be a "sealed" pack design. As the air-cooled pack is essentially an "open system," it becomes very difficult to achieve an IP69 sealing level on the pack, which makes it a very good solution for a pack that is mounted inside the vehicle or inside a building or container, but not as good a solution if it is mounted external to the vehicle.

The other common method of active thermal management is by forcing a liquid, often a 50/50 water and glycol mix similar to that used in engine cooling, through a series of plates that are mounted next to the cells. This system involves a liquid distribution system that must be integrated into the pack, most often done through a series of hoses and heat exchangers. The benefit of the liquid cooling system is that it is quite an effective medium for quickly transferring heat away from the cells. It can also, with a heating element in the system, be used to provide heated liquid in order to warm the battery in the cold weather. The disadvantages are that it tends to be a heavier system (greater mass) and there is always a risk of leaks in the battery pack. As the liquid-cooled system is essentially a "closed system," it tends to be much easier to seal to the environment making it a good solution for packs mounted external to the vehicle.

In the liquid-cooled pack, there are essentially two methods for managing the heat generation of the lithium-ion cells. In the first method, you can develop a plate that is affixed directly to the cells and flow the cooling/heating liquid directly through these plates (Figure 7).

The second method is to create a single plate through which the fluid flows, but rather than affix the cells directly to this plate, a series of "fins" are attached to the heat sync plate. The lithium-ion cells are then attached directly to these fins (Figure 8).

One other aspect to examine in actively cooling and heating lithium-ion batteries is direct and indirect cooling. These terms are sometimes used interchangeably with active and passive. Direct cooling refers to moving the cooling/heating medium directly across the cells, whereas indirect cooling/heating involves moving the cooling medium through a heat exchanger that is

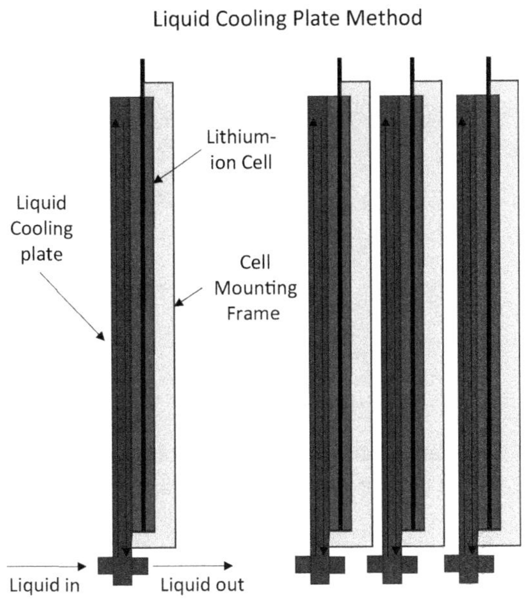

Figure 7
Liquid cooling plates.

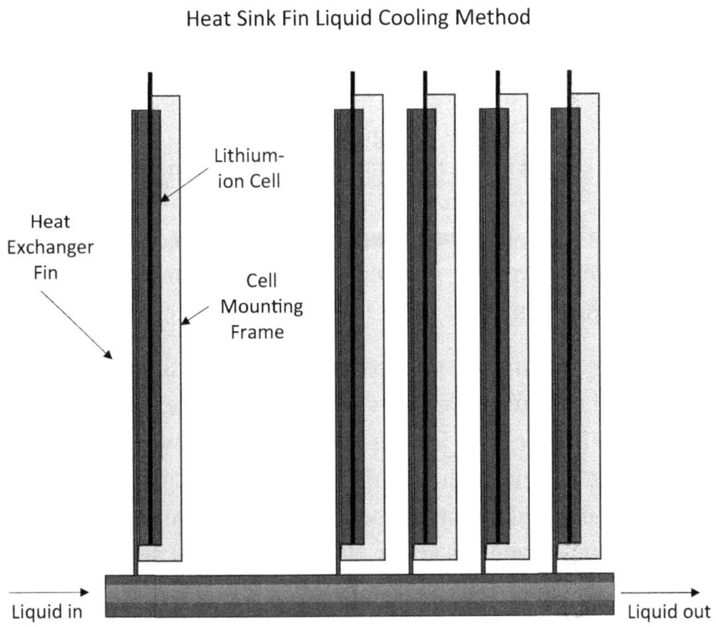

Figure 8
Heat sink fins.

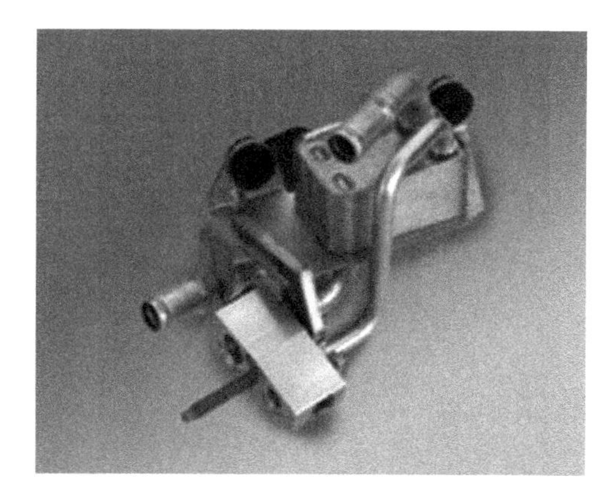

Figure 9
Behr refrigerant-based battery cooling system.
Courtesy MAHLE International GmbH.

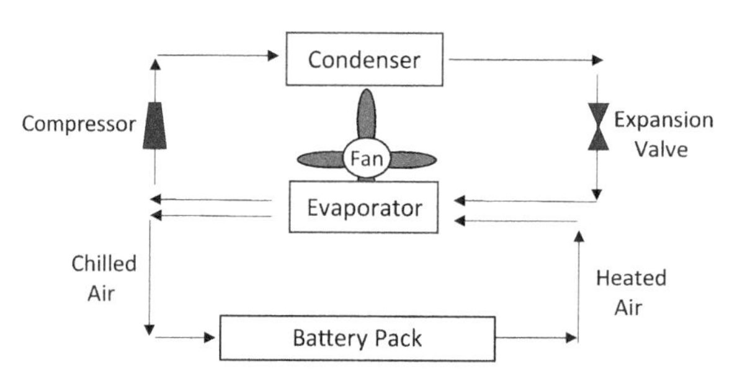

Figure 10
Direct refrigerant-based cooling schematic.

attached to the lithium-ion cells. In actual usage, direct cooling is much more feasible with an air-cooled system where the thermal system can be designed such that the air is either pushed or pulled directly across the cells.

Refrigerant-based cooling systems are yet another method for thermally managing a battery system. Refrigerant-based systems have many of the same benefits as liquid-cooled systems, but tend to have a slightly higher cost. On the positive side, they eliminate the risk of liquid leakage into the battery pack. One such example is the system introduced by the Behr Group, now MAHLE International GmbH, who introduced a refrigerant-based cooling system targeted at HEV- and EV-type batteries (Behr, 2010) (Figures 9 and 10).

Heat pipes have been discussed and evaluated for various lithium-ion-based battery thermal management systems, but as of today have not yet been able to offer either the performance or the cost advantage to make them worth investigating for large energy storage systems. In principle, the heat pipe combines two different thermodynamic principles, thermal conductivity and phase change, in order to transfer heat. The heat pipe operates by sealing a liquid into a series of metal pipes. As the pipes are heated, the liquid converts into steam and returns to the other end of the pipe where it again returns to a liquid form as it cools. Heat pipes have found significant market penetration in the portable computer market but have yet to find a home in automotive or large energy storage systems. The biggest downfall of the heat pipe for automotive applications is that above a certain temperature (depending on the liquid) all of the liquid is converted to steam and the system thereby stops transferring heat at all.

Another thermal management device that has been evaluated by some designers is the Peltier device. The Peltier device is a nonmechanical device that is made up of two different types of materials. When current is applied to one side of the device, heat is transferred from one side to the other where a heat sync can be used to move the heat out of the pack. This can be used to create what I refer to as a hybrid thermal management system. A hybrid thermal management system is one that may include various combinations of liquid, air, and other cooling mediums combined into a single battery pack system. One example may be a system design that seals the pack or module with an internal fan that circulates air within the pack; this would be combined with a heat exchange device such as an aluminum manifold or a Peltier device that the air is passed across. A cooled liquid that is entirely outside of the pack could then be run through the heat exchange device thereby cooling the air that is being circulated within the pack.

One more thing that must be considered in your thermal management system design is the pressure drop. The pressure drop is the difference between incoming and outgoing pressure of the fluid, which can be either liquid or air, and is basically a measure of the amount of pressure that the system must overcome in order to flow through the cooling system. In a liquid-cooled system, the pressure drop will determine the size and flow rate of the pump that is needed to move the fluid through the cooling system. In an air-cooled system, it will help to determine the size and flow rate of the fans that are needed to move the air through the system.

Passive Thermal Management Systems

Passive thermal management is the process of managing the temperature of the cells and the pack without forcing air, liquid, or other cooling medium into the pack. This can be accomplished through several methods. Most frequently, this is achieved through the use of aluminum or metal housings transferring the heat of the cells to an aluminum or metal pack enclosure. This allows the heat to be dispersed throughout all of the metal of the pack and enables it to radiate into the environment. This is generally referred to as a heat sync and is most effective on packs with low discharge rates as they generate less heat. Another method of passive cooling is to design the pack enclosure with fins such that as the vehicle is moving, air is forced over these fins thereby cooling the pack.

Figure 11
AllCell Phase Change Composite (PCC™) material.

Another method of passive cooling is through the use of a phase change material (PCM). A PCM is a material that will go through multiple physical phases when heated, generally from a solid to a liquid. That is a bit of a misnomer however as the PCM does not actually become liquid, but rather it would be more accurate to say that it softens. In these designs, the PCM is generally a block of solid material, often based on wax and graphite, which is machined or molded with locations to insert the cells. As the cells heat up, the PCM absorbs the heat and disperses it throughout all the PCM causing the material to soften or melt. The PCM is an effective cooling methodology as it takes a lot of energy to force a material to go through a phase change. This can be a cost-effective solution, but it is still limited to being able to top out at a maximum temperature beyond which it will entirely melt and become ineffective as a cooling agent (Figure 11).

The biggest benefit of passive thermal management is in its cost, as there is no additional hardware required for cooling the system and cost will be lower than an air-cooled or liquid-cooled thermal management system. The disadvantages are that it is slow to respond to high-demand applications and it is limited in its capacity to cool down only to the ambient air temperature.

Temperature—Protection and Insulation

Another aspect of your thermal management system is the integration of heat shields in strategic areas in order to deflect heat away from both external and internal components. Depending on the ESS design, installation location, and amount of heat generation, you may need to include some sheet steel or aluminum to protect against radiant heat and to direct it away from the lithium-ion cells. There are also some more advanced heat shields that use a sandwich-type design where two pieces of sheet metal are separated by an insulator material in order to isolate the heat away from specific components of the system.

In some systems, the battery enclosure itself may act in this manner and no additional shielding may be necessary. Another method that is frequently used for keeping the heat either in or out of the battery enclosure is the use of insulation. In these systems, the insulation is used for two reasons. In the summer time and in hot ambient environments, it may be used to protect the battery from the high ambient temperatures. And in cold regions or in winter usage, it may be used to avoid heat loss to the outside environment. But even in these systems, it may be necessary to shield the electronics from the batteries.

Thermocouples and Measurement

Temperature measurement is one of the most challenging issues in thermal system design and management. The most accurate location to measure the temperature of a lithium-ion cell is actually inside the cell. However, it is also the most difficult location to install a temperature sensor unless it is designed into the cell itself. Therefore, typical cell measurements are taken from the surface of the cell. When outfitting a cell for testing, many thermal engineers will install temperature sensors in three locations on the cell: near to the terminals, on the middle of the surface of the cell, and on the surface farthest away from the terminals. This is usually only done during characterization of the cell however, as the cost of installing these many thermistors in a large pack would be very cost prohibitive.

For these reasons, production designs will usually include two to three temperature sensors per module. Installation of the temperature sensors in this case is not dissimilar to how the sensors are located during cell characterization. One temperature sensor may be mounted on one of the first cells in the module, one in the middle, and one near the end. Of course, the proper method to determine exactly how many sensors are required and where to locate them should be based on your module characterization testing and CFD analysis, both of which will inform you where the "hot" spots in the module design are.

Remember that the purpose of the temperature sensors is to identify and manage the overall temperature of the pack, not necessarily to monitor the temperature of each cell. The temperature sensors in the thermal management system (TMS) are intended to monitor the temperature of the pack and inform the BMS when to send more heating or cooling to the battery.

In the thermal system design, for both liquid- and air-cooled designs, it is also good practice to install temperature sensors to monitor the temperature of the incoming fluid (liquid or air) as well as the temperature of the outgoing fluid (liquid or air). Note that this is in addition to the temperature sensors that are located inside the battery modules.

Another important concept to remember is the difference between outside ambient temperatures and the internal energy storage system temperatures. Many specifications will describe a temperature range for a system as ranging from –40 °C to 70 °C, but this is not referring to the temperature that the battery will achieve. Only the external temperatures that the battery must be designed to operate within. However, the thermal management system must still be designed to keep the lithium-ion cells operating within their optimal range. So for the higher temperature climates, chilled air or refrigerant-based cooling system may be needed in order to reduce the cells down to the 23 °C range. And in colder climates, a heater may be needed in order to provide warmed air or liquid to the battery in order to bring it up to its operating range.

Two of the main types of temperature-measuring thermocouples that are used in battery systems today are thermistors and negative thermal coefficients (NTC). A thermistor is a device that measures temperature through the use of two dissimilar metals that contact each other in one or more locations. When a temperature difference is experienced in one of the two conductive metals, a voltage is produced. This actually operates using the same principal as that of the Peltier device, which was discussed earlier. Thermistors operate over a very wide range depending on the type and selected from −180 °C and all the over 1800 °C, but typical measurement range is from 0 °C to over 1100 °C.

The negative thermal coefficient (NTC) thermistor is a temperature measurement device that operates by measuring a decrease in electrical resistance as the temperature increases. NTC thermistors operate in the range of −50 °C to 150 °C but can operate up to about 300 °C for some of the glass-encapsulated NTC units. Finally, thermocouple selection will depend on the desired temperature range, tolerance, cost, stability, and output.

Design Guidelines and Best Practices

- Lithium-ion cells like to be maintained at about the same temperature range that people are comfortable with, which is about 23 °C (73 °F).
- Thermal management of a battery is the integration of some form of heat exchange device into the battery system in order to maintain the battery cells at a near constant temperature.
- Thermal management system will be dependent on three factors: the duty cycle that the pack will operate under, the region where it will be used in, and the cell itself.
- Three different types of heat transfer need to be considered: conduction, convection, and radiation.
- Active thermal management involves using some medium such as air, liquid, or refrigerant that is forced through the pack and over the cells to reduce temperatures.
- Passive thermal management is the process of managing the temperature of the cells and the pack *without* forcing air, liquid, or other cooling medium into the pack.

CHAPTER 11

Mechanical Packaging and Material Selection

In this chapter, we will discuss how to do a "back of the napkin" estimate of a battery pack's physical size and volume. We will also look at various types of packaging materials and solutions, ingress sealing protection considerations, shock and vibration considerations.

Perhaps the first and most important statement we can make about battery packaging is this: there is no standard size lithium-ion battery pack and there is not likely to be one in the near future. This is a question that is quite frequently asked but the reason that there is not likely to be any standardization around battery packs is relatively simple: each battery pack is designed to fit one or two different vehicle architectures for a single manufacturer. Every vehicle manufacturer has different architectures, different ways that the vehicles go together, different places they mount, and install different component. In larger stationary systems, there is a direction to use standard 19-inch-server racks to install lithium-ion batteries into. However, some battery manufacturers have developed different energy storage system (ESS) mounting solutions. Because of these issues standardization at the module or pack level is not all that likely… at least not in the near term.

A secondary reason that standardization is a challenge is that almost every electrified vehicle on the market today is a "retrofit." In other words, it is a vehicle that was designed for a gasoline or diesel internal combustion engines where the manufacturers are trying to find existing space in which to package a battery system, rather than a vehicle that was designed from the ground up, to be an electrically powered vehicle. This means that the batteries are being mounted in places that in a ground up design they likely would not be mounted—in the trunk, under seats, in the transmission tunnel, in the fuel tank location.

While this is the state of the automotive industry today, there will come a time when we move to more purpose built applications. Vehicles tend to go 10 to 15 years between "all-new" architecture redesigns. This means that we are probably two to three battery product generations ahead of a time when vehicle architectures are designed with electrification in mind. This means designing the battery in, instead of trying to package it in an existing space. Today, there are only a few manufacturers, such as Tesla, who offer electrified vehicles that are ground up designs, but more will move this way over time.

As we begin to look at the mechanical aspects of lithium-ion ESS, it is usually best to begin with the requirements. What is the application? Where will it be mounted or installed? If an

The Handbook of Lithium-Ion Battery Pack Design. http://dx.doi.org/10.1016/B978-0-12-801456-1.00011-7

automotive application, will it mount under the hood, in the passenger compartment, under the chassis, in the trunk? If stationary power application, will it be in a mobile enclosure or will it be permanently mounted? If a marine application, what sealing and environmental requirements will it have? Once we understand the location for the system installation, we can begin looking at the other requirements. Does it need to stand up against harsh environmental requirements? What level of sealing does it need to have? Is it a structural component in the system? What kind of shock and vibration will the system experience?

Next we can evaluate the other mechanical needs. Does it mount in a crash zone within the vehicle? Does it need to meet any knee loading requirements, in other words does it need to be able to stand up to someone standing or stepping on it? Additionally, we must evaluate other influences such as electromagnetic interference and electromagnetic compatibility impacts, the distance to other heat generating components in the vehicle, the distance to the passengers and the need for serviceability of the pack.

The components of the ESS that we are interested in here from a mechanical and structural perspective include the housing or enclosure within which ESS is installed and the mechanical structure that encloses and protects the cells, referred to as the module. Both of these components may include a combination of steel, aluminum, plastic, fiber glass, and composite materials.

Understanding the level of shock and vibration that your ESS will experience is also a key aspect of your mechanical design. For instance, in automotive applications a high level of vibration resistance may be necessary. This may be accomplished through a combination of design and material selection. In some cases, the designer may include a foam type of material in order to reduce the impact of the vibration on the battery pack. It is also vital that the flow of the vibration through the battery does not go through the cell and the cell interconnections. The mechanical system design must account for this and isolate the cells and their connectors since this will cause increasing impedance, increasing levels of heat, and eventually premature failure. For other systems such as may be integrated into very large systems, the impact of shock on the battery must be evaluated and accounted for in the design.

Module Designs

Since we are talking about mechanical and structural components of the battery system, let us take just a moment to discuss the different types of battery modules that are in use today. The term battery module is generally used to refer to the assembly of lithium-ion cells into a single mechanical and electrical unit. The module consists of the lithium-ion cells, bus bars, voltage/temperature monitoring printed circuit board, thermal management components, and finally the overall mechanical structure.

The module design will depend highly on a couple of things. First is the type of cell being used as this will necessitate the final design configuration. For instance, the use of a pouch-type cell will require a set of frames made from either plastic or metal that the cells are mounted to and in which the cells are generally protected and the appropriate amount of pressure is applied onto the cells. For the large prismatic can cells, each cell may not need its only frame/structural piece. In this instance, the interconnect board (ICB) may be enough to provide the appropriate amount of mechanical structure.

In some instances, the use of banding is included in order to assemble the complete module design. However, there are a couple of long-term issues that need to be evaluated when it comes to the use of banding your modules. If you are using a plastic-based banding, the question comes down to what is the level of elasticity over the life of the material? If the banding material stretches enough over time, the cells may no longer get the amount of stack pressure that they need. If steel banding is used, it may have the opposite effect in that it will not stretch over time, but the cells will grow dimensionally over their life. Therefore, the amount of pressure that the metal banding provides will increase over time.

Another thought that should be considered in a module design is the need for serviceability. Some battery manufacturers have designed their modules in such a manner that the cell interconnections are made with mechanical components, bolts, and nuts. In this case, the cells can be replaced and the module can be serviced throughout the life of the module. However, there is always a risk that the mechanical connections will loosen over time thereby increasing the resistance and creating the potential for a failure mode in the battery. Other battery manufacturers believe more strongly in the use of welding for the cell-to-cell interconnections. The benefit here is a somewhat lower material cost (no costs for fasteners) and somewhat higher reliability (will not loosen over time). However, there is also the risk that if you build a module with multiple cells and there is a failure of one cell, the entire module must be replaced instead of a single cell.

For most battery manufacturers, the module is the basic building block for all of their system designs, so having a solid and reliable modular solution can be a very big benefit to driving commonality across multiple systems.

Use of Metals in Battery Design

We will begin by looking at the different types of materials that are used in battery enclosures. Module and pack enclosures may be designed from plastics, steel, aluminum, fiber glass, or composite materials. And in almost all cases, a combination of these materials will be used in an ESS design.

First, we will review the use of a stamped steel enclosure. Steel offers many benefits including high strength and relatively low cost. However, the use of a steel enclosure must also generally include some weldments or otherwise mounting attachments and may need to add mounting structures and to provide for strength. This adds to the processing time of the material and increases the cost. Several companies are today working on methods of increasing the strength and reducing the weight and mass of steel in order to make it more competitive with low-weight aluminum applications. The Nanosteel Company is one such company who offers advanced steel solutions that begin to achieve similar weight as aluminum but still retaining the high strength of steel (Nanosteel, 2014).

Aluminum enclosures can either be of stamped or die-cast methods. Aluminum offers lower weight than steel but can require additional material thickness in order to meet strength requirements, especially in stamped pieces. The other method that can be used with aluminum is in the form of die casting. This can be either through high-pressure die casting (HPDC) which offers the best strength, porosity, and surface quality but can be very expensive to tool. Sand casting tends to be less expensive to tool but the part quality is often so low that additional finishing steps may be required. Plaster casting offers the best of both worlds, it is relatively low cost to tool and offers finishes nearly as good as the HPDC methods. The biggest challenges with plaster casting are the porosity of the flow which can create weak spots in the final product. That makes plaster casting, at least for large pieces, most beneficial as a prototype solution.

HPDC aluminum offers the benefit of being able to design a part with many features integrated into the cast itself, from mounting features to air flow channels, to structural ribs, and supports. This in addition to its low weight makes it an ideal solution for many lithium-ion battery solutions.

If the system design is based on one or more of these types of metals, then it is also important to evaluate the need for coatings. There are several reasons that you may want to use a coating and even several types of coatings. One reason to add a coating is to prevent grounding and shorting of the electronics and battery cells, in other words, to add an isolation function to the battery system. This may be available in one of several materials and types. One common usage is to use an isolating film that has an adhesive on one side. This can then be installed directly to the metal to provide the level of isolation needed. Another method for coatings is to provide either a liquid coating or a powder coating. These two methods are more often used to provide environmental protection rather than isolation.

When using metal enclosures, the design must be evaluated for strength. It is often necessary to include structural reinforcements in the enclosure. This may take the form of either additional metal components that are bolted or welded into plate, or for cast enclosures it may take the form of adding ribbing into the design. Determining which and how much reinforcement should be added should be managed through finite element analysis structural, shock and vibration management.

Special care must be taken when dissimilar materials are used together as some dissimilar metals, when used together, can create unwanted chemical reaction and cause one of the metals to corrode. This is called galvanic corrosion and is especially important in electrical systems as it is the result of the electrical charge and can cause long-term reliability issues as the metal corrodes over time.

Use of Plastics and Composites in Battery Design

Some smaller battery systems may use plastic enclosures. These systems are more likely to use a plastic if the battery does not have major structural demands. In some of the hybrid and stop-/start-type automotive batteries, the enclosure only needs to enclose and protect the lithium-ion cells with minimum structural demands. Internal to even large battery pack, plastics and polymers are used in multiple areas. For some of the larger automotive ESSs, a composite cover may be used in concert with a metal base. For instance, General Motors Chevrolet Volt uses a sheet molded composite made of a lightweight vinyl ester resin that contains haydite nanoclay filler and 40% glass fibers (Vink, 2012). In simpler terms, the cover for the Chevrolet Volt is made of fiber glass.

The Volt and its sister vehicle the Ampera use lithium-ion pouch-type cells and so also use a plastic "end" and "repeating frames" to separate each of the cells in their systems. These are injection molded plastic pieces made from BASF's nylon 6/6 grade and Ultramid 1503-2F NAT, which is 33% glass filled and hydrolysis stabilized (LeGault, 2013).

Another component that is frequently designed with plastics is the ICB. The ICB is often a plastic piece that integrates the cell interconnects, wiring harnesses for temperature and voltage monitoring, electronic monitoring circuitry, mechanical support for the cells, and cell vent management. Due to the amount of functions, this may take the form of an over-molded plastic piece with copper or nickel coated copper bus bars. The GM Volt ICBs use DuPont's Zytel 7335F grade of PA6 for the base and connector housing, and a 35% glass filled PA66 by Zytel for over molding the ICB (Vink, 2012).

The other topic that should be considered in respect to plastics and polymers used in batteries is the flame retardant rating. Flame retardant ratings range from V0 to V2, the V stands for vertical flame rating but there are also horizontal flame ratings both of these sets of flame retardant ratings are based on the Underwriter's Laboratory (UL) UL-94 standard (UL Prospector, 2014). It is important to ensure that the plastics used in your ESS design are nonflammable. The reason for this is that in the event of a thermal runaway event of one or more lithium-ion cells you do not want to add more fuel to the combustion process. So if your plastics are V0 rated they will be less susceptible to the initial failure thereby making it easier to manage the cell failure.

Sealed Enclosures

ESS enclosure design must take into account the usage and installation location of the battery system. In virtually all battery systems, it is important to evaluate the international protection

(Table 1), or ingress protection (IP), rating needs of the system. For automotive systems, it is most frequently desired to ensure that the system is sealed against both liquid and dust intrusion, which is known as IP6K9 rating. The IP rating code was developed as part of the International Electrotechnical Commission (IEC) standard 60529. The first digit represents the level of protection that the unit provides for dust and physical intrusion, in other words it protects against dust getting into the pack as well as protecting against getting things like fingers into your battery pack design. The second character refers to the protection against liquid intrusion ranging from dripping liquids to full submersion and water spray (Maxim Integrated, 2007).

Another type of enclosure rating that must be thought through is the National Electrical Manufacturers Association (NEMA) rating system (Table 2). The NEMA rating system was set out for electrical equipment that is intended for installation in either indoor or outdoor systems. Therefore, it is more common to find in stationary and grid-type applications, both industrial and household. When it comes to NEMA-rated enclosures, there are many manufacturers that currently produce standardized NEMA enclosures. So for smaller ESSs, it is easier to purchase an off-the-shelf enclosure that already meets the NEMA standards. The NEMA and IEC IP standards are similar, but NEMA ratings and testing specify the degrees of protection against mechanical damage of equipment, risk of explosions, or conditions such as moisture (produced, for example, by condensation), corrosive vapors, fungus, and vermin as compared to the IEC IP which only indicates protection against dust and liquid intrusion (National Electrical Manufacturers Association, 2005).

Table 1: International protection classes

Level	Dust and object protection	Level	Liquid protection
0	No protection	0	No protection
1	Protection against body surface touch	1	Protection against dripping water
2	Protection against finger insertion	2	Protection against dripping water at 15° tilt from normal position
3	Protection against tools and thick wires	3	Protection against spraying water
4	Protection against small parts—wires and screws	4	Protection against splashing water
5	Dust ingress not entirely protected, no contact	5	Protection against water jets at 30 kPa
6	No ingress of dust, no contact	6	Protection against powerful water jets at 100 kPa
		6K	Protection against powerful water jets at 1000 kPa
		7	Protection from immersion up to 1 m
		8	Protection from immersion beyond 1 m
		9K	Protection against powerful high-temperature water jets

Table 2: National Electrical Manufacturers Association enclosure rating system

	Enclosure Rating									
Protection against	1	2	4	4X	5	6	6P	12	12K	13
Access to hazardous components	X	X	X	X	X	X	X	X	X	X
Ingress of solid foreign objects (falling dirt)	X	X	X	X	X	X	X	X	X	X
Ingress of water (dripping and light splashing)		X	X	X	X	X	X	X	X	X
Ingress of solid foreign objects (dust, lint, fibers)			X	X		X	X	X	X	X
Ingress of water (hose down and splashing)			X	X		X	X			
Oil and coolant seepage								X	X	X
Oil and coolant spray and splashing										X
Corrosive agents				X			X			
Ingress of water (occasional temporary submersion)						X	X			
Ingress of water (occasional prolonged submersion)							X			

Design Guidelines and Best Practices

- There is no standard size lithium-ion battery pack
- The term battery module refers to the assembly of lithium-ion cells into a single mechanical and electrical unit
- The module is the basic building block for all system designs
- Battery enclosures may be made of a combination of stamped steel or aluminum, cast aluminum, fiber glass, composite, or plastics
- Plastics are commonly used in battery designs in the ICBs, modules, thermal systems, and in the module and pack mechanical structure
- The two types of sealing standards that must be assessed are IP and NEMA

CHAPTER 12

Battery Abuse Tolerance

Now that we have reviewed all of the major subsystems in the lithium-ion battery, we need to discuss testing and abuse of the battery system. In all cases, the battery system is designed to optimize the performance and protect the lithium-ion cells. However, there are occurrences that drive the cells out of their normal operating range. Most of these have been discussed in relation to the design guidelines around them, but now we will discuss in relation to the Design, Validation Plan & Report (DVP&R) planning and testing that must be conducted in order to successfully ensure that the cells safely operate within their predetermined range.

Lithium-ion battery testing generally falls into one of three different categories: (1) characterization and performance testing, (2) abusive testing, and (3) certification testing. Characterization and performance testing is focused on evaluating how the battery will perform under a specific set of testing criteria. Characterization testing is most often conducted at the cell level in order to understand the basic performance of a cell and at the module, pack, or system level in order to evaluate how the pack will operate during specific performance profiles. Abuse testing involves forcing the battery into situations that will drive it to fail and evaluating the results in order to make changes in the design to mitigate these failures. Abuse testing may include overcharge testing, overvoltage testing, nail penetration testing, short-circuit testing, and even drop testing. Abuse testing is done in order to find the limits of safe performance of the cells and packs. Finally, certification testing is done by completing a very specific set of tests that are determined by the relevant certification body and have various focuses. Examples of certification testing include tests such as the United Nations (UN) testing guidelines for the transport of hazardous goods in order to certify that the product can be shipped via public or private air or road transportation.

During the design process and throughout the ongoing discussions with the customer, the battery test engineer should gather enough information to begin developing a DVP&R plan. The DVP&R plan will include the specifics of all testing to be completed, including the quantity of units to be tested, the duration of the test, the location of the test, and the pass/fail criteria for each of the three types of testing.

In the sample report below (Table 1), you see an example of what a DVP&R plan may look like. The DVP&R plan should include all of the tests required to be completed and link them to a specific set of requirements or certification criteria as well as the criteria that will indicate that the test was passed. The plan should also include a section on the test report to ensure that all test results are accurately documented. This document becomes a living tool that is

The Handbook of Lithium-Ion Battery Pack Design. http://dx.doi.org/10.1016/B978-0-12-801456-1.00012-9

Table 1: Design, Verify, Plan & Report example

Design Validation and Verification Plan and Report												
Test Plan										Test Report		
Test #	Specification & Test Method	Test Description	Acceptance Criteria	Test Stage	Target Requirements	Test Responsibility	Test Start	Test Complete	Sample Quantity	Pass/Fail	Actual Results	Notes

continually updated as the test results begin to come to completion. It is typical to create a table such as this one for each phase of the design and testing process.

Failure Modes of Lithium-Ion Batteries

Perhaps at this point it would be valuable to briefly discuss the manners in which lithium-ion batteries *could* fail before we begin discussing the actual types of testing that may be done. We can generally break the failure modes down into two different categories: internal failure modes and external failure modes.

An internal failure mode can occur from a manufacturing defect, such as manufacturing debris getting inside the jellyroll. Another internal failure that may occur is an increase in internal resistance, which generally may occur toward the end of the batteries life. An external failure mode can occur when the battery is either operated outside of its safety zone or due to a failure of the controls or thermal systems. The worst case external failure mode occurs during a vehicular collision or when one or more cells go into thermal runaway.

One of the biggest lithium-ion battery recalls that has occurred in the past 20 years was at least in part the result of an internal failure mode. In the early 2000s, Sony recalled hundreds of thousands of lithium-ion batteries that were used in laptop computers due to the risk of fire. While there may have been multiple potential causes for the failures of these cells, one item that was identified was a flaw in the manufacturing process when the lids of the 18650-type cells were crimped in place. Upon investigation it was found that the nickel plating on the cans was being damaged and flaking off micro-sized particles during the crimping process. This resulted in material entering the jellyroll and ultimately with an internal short circuit (Sony, 2006). Subsequently to this, Sony among many other manufacturers changed their assembly processes to move away from crimping and toward laser welding the lids in place.

Another, and more common, internal failure mode is the increase in internal resistance of the cell. By measuring the rate at which the internal impedance (resistance) grows, it is possible to determine the life of the cell. However, this is not as easy as it sounds. The internal impedance is essentially measuring the growth of the solid electrolyte interphase (SEI) layer on the anode. As the cell cycles, lithium-ions tend to get trapped in the SEI layer over time. And as more and more lithium-ions get trapped inside the SEI layer, it becomes more difficult for the free lithium-ions to pass through to the cathode (Voelker, 2014).

Other internal failure modes may include a chemical breakdown of the electrolyte, which could include generation of gasses within the cell. This generally occurs when the lithium-ion battery is being used at higher temperatures. In fact, many cell manufacturers include special additives in their electrolytes in order to prevent the cell from going into premature thermal runaway. These additives will generally begin to create gasses within the cell at certain temperatures.

The separator may also begin to break down at high temperatures, depending on the type of separator, which causes the flow of lithium-ions to slow and eventually stop as the "pores" in the separator become melted or clogged. Again this tends to happen at high temperatures and is in fact designed to begin failure modes at certain temperatures in order to prevent thermal runaway events from occurring.

Yet one more internal failure mode that can occur is called lithium plating, or in some areas it is called "white out." This condition occurs when the cell goes into overvoltage or undervoltage operating conditions. In these conditions, the lithium-ions will get stuck, typically in the anode, and will no longer pass back and forth in the cell. The more lithium that gets stuck the more plating occurs and the more internal resistance grows and capacity fades.

External failure modes range from operating the batteries at continuously high loads or high temperatures and less frequently from impacts and vehicle crash scenarios. MIT's Impact Crashworthiness Laboratory has been working to understand the causes and possible solutions and remedies for lithium-ion battery design under impact conditions (Sahraie, Meier, & Wierzbicki, 2014; Xia, Wierzbicki, Sahraei, & Zhang, 2014). When lithium-ion cells are punctured during an impact an external short-circuit event is the most likely scenario. This type of external failure mode causes a rapid loss of electrolyte, massive heat generation, and under most conditions a thermal runaway event. The MIT group has proposed the use of increased strength metals to "armor plate" the bottom of the pack in order to reduce the potential for external debris to penetrate into the pack. This of course is assuming that the battery pack is mounted external to the vehicle, underneath it rather than inside the vehicle. However, even with a battery that is mounted internally there is a risk of penetration during a crash event. Many automotive original equipment manufacturers (OEMs) will conduct crash simulations to ensure that the battery is located out of the "crush zone" of the vehicle. Some are even going as far as designing the battery to become a structural member of the vehicle.

Characterization and Performance Testing

Battery performance testing typically occurs in several phases, beginning with cell characterization. The purpose of cell characterization is to determine how the cell will perform under a certain set of operating criteria. This allows the system engineer to design a system that ensures that the cells do not operate outside of this range. Characterization is usually a series of charge/discharge cycles that are performed at multiple temperature ranges under identical test cycles.

Characterization testing is generally conducted using a "cycler" or "channel" that provides current and power to the cell, module, or pack, which is controlled with a separate unit which provides the programmable characteristics (Figure 1) and finally is done in a thermal chamber to manage the temperatures the cell will experience during the testing. The thermal chamber

Figure 1
Arbin BT 2000 battery tester.

enables the cells, modules, or packs to be tested at specific temperatures generally from −40 °C up to about 60 °C and above (Figure 2).

In addition to the performance testing, there is also usually a certain amount of life cycle testing that must be done during the characterization stage. The purpose of this is to ensure that the battery will meet the required power and energy at the end of its useful life as well as to ensure that the battery can meet its warranty targets. An example of a characterization testing suite could include tests such as cycle testing is perhaps the longest running portion of the characterization testing as it can take over a year to complete, depending on the size of the battery. For example, it is not unheard of for a cycle life test for a large plug-in hybrid electric vehicle (PHEV) or battery electric vehicle (BEV) battery of between 10 kWh and 16 kWh to take in the range of 400–420 days to complete due to the size of the usable energy that must be fully discharged and charged (cycled) along with whatever rest period is required between discharging and charging the battery pack. In this example, you may be lucky to achieve one cycle per day. For example, if were testing a Chevy Volt battery with 10.5 kWh of usable energy at 120 V and 15 A (about a 1° C rate) it would take between 3 and 4 h to fully charge the battery and then the same again to discharge it. And if the life of the battery assumes one cycle per day for 10 years, that is 365 cycles per year

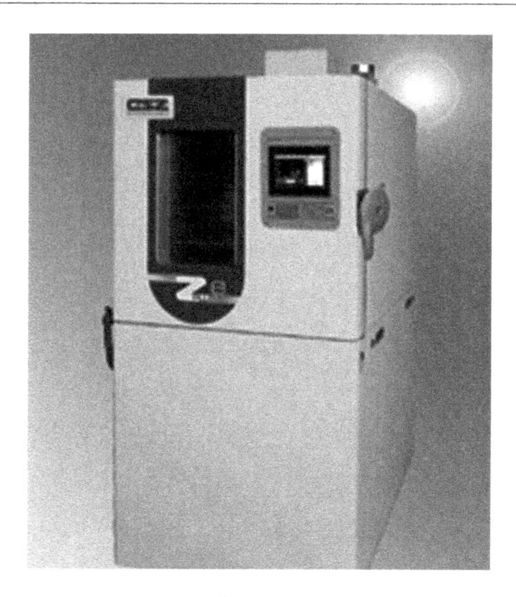

Figure 2
Cincinnati Sub-Zero thermal chamber.

times 10 years, or 3650 cycles. So even at three cycles per day it would take over 1200 days (over 3 years) to complete this test schedule!

One the other side of the coin, if you are testing a single cell, especially if it is a relatively small one, you may be able to achieve multiple cycles in a day. Keep this in mind as you are developing your DVP&R plan as the cycle life must generally be completed using the final production intent design instead of an early prototype level. That means that the cycle life testing may run right up to start of production depending on your testing and development plan.

Another example of a "standard" characterization test is the testing outlined in the United States Advanced Battery Consortium LLC program "FreedomCAR Battery Test Manual for Power-Assist Hybrid Electric Vehicles" (2003). The testing described in the FreedomCAR manual is focused on performance-type testing and includes:

- Static capacity testing—the purpose of this testing is to define the battery capacity in both ampere hours and in watt hours for a battery at a specified discharge rate.
- Capacity fade—shows the amount of irreversible capacity loss that a battery will experience based on the above static capacity testing.
- Hybrid power pulse characterization (HPPC) testing—this is perhaps the most well defined in the FreedomCAR manual. The HPPC testing is used to compare the cell or pack performance against a set of performance goals (such as the FreedomCAR goals). The HPPC test profile is a relatively simple one, it includes a short period of discharge, followed by a rest period then a short period of charging (regenerative braking). This is then repeated for a period of time (Figure 3).

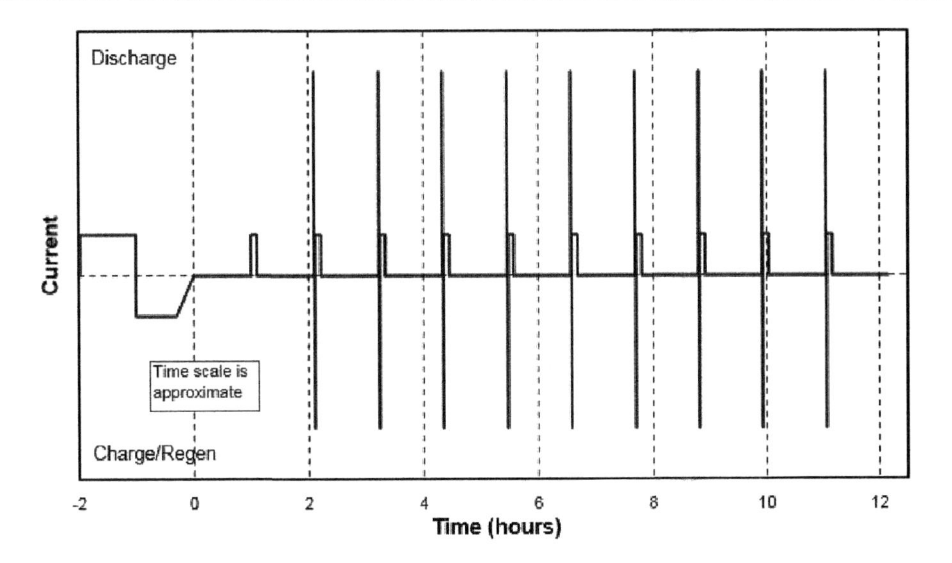

Figure 3
Typical HPPC charge/discharge testing cycle. *FreedomCAR Program Electrochemical Energy Storage Team (2003).*

From this testing, a mountain of additional data becomes available, including the resistance as a function of the depth of discharge, the pulse power capability, the amount of energy available at different depth of discharge, the available power at different depth of discharges, how much power and energy fade the battery will experience over its life, the maximum and minimum depth of discharge values the battery is capable of achieving and still meeting the performance requirements, and the amount of heat generated during operation.

Additional characterization testing that may be included in your test plan may include:

- Self-discharge tests—help the battery designer to determine how much energy is lost over time as the battery sits in storage.
- Cold-cranking tests—show how much power the battery can provide at low temperatures to "crank" the vehicles engine and restart it.
- Thermal performance testing—describes how much power and energy are available at different temperatures.
- Energy efficiency test—describes the "round trip energy efficiency" of a battery cell or system. In essence, this tells us how efficient the overall system design is and how much energy is lost during use.
- Cycle life testing—determines how many full charge/discharge cycles at a predefined set of operating conditions a battery will achieve.
- Calendar life testing—as it is unrealistic to actually test a battery for its full expected life, which could be 10–15 years or more, it is important to get some idea of how long in calendar years the battery will last. That is exactly the purpose of this test (FreedomCAR Program Electrochemical Energy Storage Team, 2003).

This is of course only one example of a characterization testing regime, depending on the application, industry, and customer you may end up with something very similar or something entirely different.

Another test regime that is beginning to get some use in the lithium-ion world is the starter battery assessment test for lead acid SLI batteries. While this test set is intended for lead acid starter, lighting ignition batteries, it is also very useful in testing stop/start batteries as they will work under very similar operational cycles.

Safety and Abuse Testing

Safety and abuse testing can cover a wide variety of conditions depending on the application specific requirements ranging from overcharging a single cell to overcharging a complete pack, to inserting a nail into a cell or pack to evaluate how it fails, to conducting shock and vibration testing to see how the battery fatigues over time, to spraying the enclosure with a corrosive salt spray and gravel sprays, and high-pressure water intrusion tests to replicate use in automotive applications. Not all of these tests will be required for every application so it is important to note which market, application and geographical region your product is intended to operate in in order to determine the right set of safety and abuse tests.

Published by the U.S. Sandia National Laboratories, the *FreedomCAR Electrical Energy Storage System Abuse Test Manual for Electric and Hybrid Electric Vehicle Applications* (2006) offers a good outline of what abuse testing should be conducted on an automotive battery as well as laying out the testing conditions and data recording. The FreedomCAR test manual breaks the abuse testing into three main categories. They are:

- Mechanical abuse testing—which covers crush, penetration, drop, immersion, rollover, and shock testing
- Thermal abuse testing—which includes thermal stability, simulated fuel fire, high-temperature storage, rapid discharge and charge, and thermal shock cycle testing
- Electrical abuse testing—including overcharge and overvoltage, short circuit, overdischarge and voltage reversal, and partial short circuit (Sandia National Laboratories, 2006).

This document was based on the work done by the United States Advanced Battery Consortium LLC, an industry group made up of U.S. OEMs, government laboratories, and the Department of Energy, and their "Electrochemical Storage System Abuse Test Procedure Manual" (1999) which offered one of the first abuse testing manuals. Many of the tests included in the FredomCAR manual are also included in the USCAR abuse testing regimen, which includes:

- Mechanical abuse testing—shock, drop, penetration, rollover, immersion, and crush testing
- Thermal abuse testing—radiant heat, thermal stability, compromise of thermal insulation, overheat/thermal runaway, thermal shock cycling, elevated temperature storage testing, and extreme cold temperature testing

- Electrical abuse testing—short circuit, partial short circuit, overcharge, overdischarge, and AC exposure
- Electrochemical storage system vibration testing (United States Advanced Battery Consortium, 1999).

Many of the automotive OEMs in the United States and in Europe use the either the FreedomCAR Hazard Severity Level or the EUCAR Hazard Levels that was developed by the European Council for Automotive Research and Development (EUCAR) (European Council for Automotive R&D, 2014) for evaluating the results of abuse testing (Table 2). These two rating systems are identical and the FreedomCAR system was based on the EUCAR system, so in essence these two groups have a common abuse testing severity rating system. In actual usage, most other global OEMs also recognize this system for identifying the level of failure.

This rating system ranges from 0 where there is no effect on the cell or battery system to 7 which is identified as an explosion and/or rapid release of energy from the battery system. The optimal results of course would be a 0, but a level 1 or 2, passive protection has been activated and defect and damage has been identified but no venting, fire, rupture, or leakage

Table 2: EUCAR hazard ratings

Hazard Severity Level	Description	Classification Criteria and Effect
0	No effect	No effect, no loss of functionality
1	Passive protection activated	No damage or hazard, reversible loss of function. Replacement or resetting of protection device is sufficient to restore functionality
2	Defect/Damage	No hazard but damage irreversible to RESS (Rechargeable Energy Storage System); replacement or repair needed
3	Minor leakage/venting	Evidence of cell leakage or venting with RESS weight loss of <50% of electrolyte weight
4	Major leakage	Evidence of cell leakage or venting with RESS weight loss of >50% of electrolyte weight
5	Rupture	Loss of mechanical integrity of the RESS container, resulting in a release of contents. The kinetic energy of released material is not sufficient to cause physical damage external to the RESS
6	Fire or flame	Ignition and sustained combustion of flammable gas or liquid (approximately more than 1 s). Sparks are not flames
7	Explosion	Very fast release of energy sufficient to cause pressure waves and/or projectiles that may cause considerable structural and/or bodily damage, depending on the size of the RESS. The kinetic energy of flying debris from the RESS may be sufficient to cause damage as well.

FreedomCAR Program Electrochemical Energy Storage Team (2003).

occurs are the design targets. However, 4 through 7 get a bit more confusing and more difficult to differentiate and generally require more additional system-level protections in order to mitigate failures in these types of events. Level 4 occurs when a cell vents and loses more than 50% of its electrolyte, Level 5 occurs when a flame or fire is detected; Level 6 occurs when there is a rupture of the cell, with or without flame, but there are no flying parts; and finally Level 7 occurs when there is an explosion, defined as flying parts and disintegration of the cell (FreedomCAR Program Electrochemical Energy Storage Team, 2003).

One other safety document that we should talk about briefly is the Pacific Northwest National Laboratory's "Inventory of Safety-related Codes and Standards for Energy Storage Systems" (Pacific Northwest National Laboratory, 2014). This is a very all-encompassing document that encompasses all laws, rules, codes, standards, and regulations relating to the development, testing, certification, and installation of energy storage systems (ESSs) for industrial, grid, stationary, and household applications. As more lithium-ion ESS installations continue to get installed the legal and regulatory environment is often still trying to catch up with the technology. This report does a very good job of evaluating the needs specific to the US market.

Certification Testing

Certification testing is conducted for two main purposes. First is to certify the product for use according to a specific industry or government application. Examples of this type of testing would include DNV certification for marine-based packs, NHTSA testing for automotive applications, and UL testing for household appliances and applications. The second type of certification testing is done to ensure the safety of people and equipment while the product is being shipped. This mostly falls under the United Nations (UN) Recommendations on the Transport of Dangerous Goods: Manual of Tests and Criteria (United Nations, 2009) with which most countries have aligned their transportation rules.

United Nations

Section 38.3 of the United Nations Handbook on the Transport of Dangerous Goods is titled *Lithium metal and lithium-ion batteries* and covers the testing that is required to be conducted *prior* to a lithium-ion cell or battery pack being shipped. UN 38.3 describes a very specific set of testing that must be conducted via an accredited test source in order to obtain approval to ship the battery. This section of the test manual was developed after several major incidents occurred on public air transport when lithium-ion cells were either damaged during shipping or some other failure occurred that drove the cells to fail while they were on a plane. Unfortunately, several incidents like this occurred in the early 2000s that drove the need for this type of certification.

It is important to note that the U.S. Department of Transportation (U.S. DOT) and most other countries recognize this as certification as the only approved certification in order to ship. However, it is not entirely globally recognized and some countries may require that the testing be done "in country" if the battery is being manufactured there.

The specific UN testing regime is different depending on the size of a battery (either a cell, module, or a complete pack) but generally follows the following eight tests:

Test T.1: Altitude Simulation	Test simulates air transport under low-pressure conditions
Test T.2: Thermal test	Test assesses the cell and battery seal integrity and internal electrical connections using rapid and extreme temperature changes
Test T.3: Vibration	Simulates vibration during transport
Test T.4: Shock	Simulates impacts during transport
Test T.5: External short circuit	Test simulates an external short circuit of the cell or pack
Test T.6: Impact	Simulates an impact
Test T.7: Overcharge	Evaluates the ability of a rechargeable battery to withstand an overcharge condition
Test T.8: Forced discharge	Tests the ability of a primary or rechargeable cell to withstand a proceed discharge condition

Not all eight of the tests need to be conducted, but each must be assessed to determine if it does need to be completed. For instance, in regard to the T7 overcharge test the UN Manual states:

> *Batteries not equipped with overcharge protection that are designed for use only in a battery assembly, which affords such protection, are not subject to the requirements of this test.*
>
> ***(United Nations, 2009, p. 397)***

The reason for this disclaimer is that much of the battery industry pushed back against this test as part of the UN certification as most all cells and small assemblies would fail this test as they are designed to be used in larger assemblies that will have the proper overcharge protections. So conducting an overcharge test on a cell that has no possible way to overcharge during transport seemed a bit unreasonable.

Similarly, as conducting the number of tests required on a very large pack—consider a PHEV, BEV, or even stationary pack that could have a cost of $10,000–$15,000 USD or more would be extremely cost prohibitive, the United Nations added another provision that if the cells and modules have all passed the UN testing, the packs do not need to again go through this testing regimen. The UN manual states:

> *When batteries that have passed all applicable tests are electrically connected to form a battery assembly in which the aggregate lithium content of all anodes, when fully charged, is more than 500 g, or in the case of a lithium-ion battery, with a Watt-hour rating of more than*

6200 Watt-hours, the battery assembly does not need to be tested if it is equipped with a system capable of monitoring the battery assembly and preventing short circuits, or over discharge between the batteries in the assembly and any overheat or overcharge of the battery assembly.

(United Nations, 2009, p. 397)

Underwriter's Laboratory

Underwriter's Laboratory (UL) has also been very active in the certification arena for lithium-ion battery testing. Most of the UL certification work and standards to date have been focused on nonautomotive applications, specifically focused on products in the mobile power, household, industrial, and commercial applications. While UL has made strides in coordinating with the International Electrotechnical Commission (IEC), the automotive OEMs have pushed back on UL certification for automotive applications. As the OEMs already require and conduct a significant amount of testing and work with organizations such as the Society for Automotive Engineers (SAE) to develop standards around battery design and testing, they have been unwilling to add another level of certification that already mimics their own but would add a new layer of cost to the already expensive battery.

To date, UL continues to participate with SAE, IEC, and other automotive industry groups but their main certification standards today are focused on portable and household power applications. However, UL has developed and issued standard 2580 "Batteries for Use in Electric Vehicles" (Underwriter's Laboratory, 2014) which offers a set of certification standards around the batteries ability to withstand simulated abuse conditions. The testing requirements of this standard mimic some of those required by the automotive manufacturers as well as the UN test manual, including a series of electrical tests, mechanical tests, and environmental tests. So the UL testing goes over and above what is required for UN transport, but has not yet been widely accepted by the major automotive OEMs as a design requirement.

The major UL certification requirements for lithium-ion batteries that exist today are:

- UL 1642 Lithium Cell
- UL 1973 Batteries for use in Light Electric Rail applications and Stationary Applications
- UL 1989 Standby Batteries
- UL 2054 Alkaline Cell or Lithium/Alkaline Packs
- UL 2271 Batteries for use in Light Electric Vehicle applications
- UL 2580 Batteries for use in Electric Vehicles
- UL/CSA/IEC 60950 (may be evaluated in conjunction with UL 2054)
- UL/CSA/IEC 60065 Batteries used in Audio and Video Equipment (Underwriter's Laboratory, 2014).

Maritime Certification

Another group of certifications that must be examined if your ESS is intended for maritime application are the DNV-GL, Lloyd's, and ABS certifications. These organizations are in the early stages of developing unique certifications for lithium-ion batteries that are used in marine applications. Today, these three organizations have a newly published set of type approval certification standards focused on the use of lithium-ion batteries at the pack or system level, but to date none of these organizations have developed cell-level type approval or certification standards.

Design Guidelines and Best Practices

- First, make sure that you understand the end use market application for the battery. That is perhaps the most important step to developing a complete and effective DVP&R plan.
- The testing plan must take into account any industry specific testing and certifications in addition to those testing requirements driven by the customer.
- Second, talk with the customer. Each customer, whether they are a major automotive OEM, a small start-up manufacturer, or somewhere in between, will have a set of testing expectations. As the battery designer, it is your job to make sure that they understand what industry and regulatory testing is required, but the testing must ensure that the customer requirements are also included.
- The three basic types of testing that are conducted on lithium-ion batteries include characterization testing, safety and abuse testing, and certification testing.
- Section 38.3 of the United Nations Handbook on the Transportation of Dangerous Goods regulates the testing that must be done in order to ship lithium-ion cells, modules, and packs.
- Marine applications require type approval and certification by agencies such as DNV-GL, Lloyd's, and ABS.

CHAPTER 13

Industrial Standards and Organizations

In this chapter, we will review some of the major industry organizations, governmental organizations, trade groups, and standards setting organizations for lithium-ion batteries. This will include both standard setting organizations such as the Society of Automotive Engineers (SAE) and the International Organization for Standardization (ISO) as well as industry groups and coalitions focused on helping the commercialization of the automotive and industrial lithium-ion battery.

The role of industry organizations is particularly important as the lithium-ion battery industry is, at least as it relates to automotive and industrial applications, still relatively young. In a young industry where the technology is still developing trade groups and industry organizations are essential as they have the ability to bring together the different areas of the value chain in one group, organization, or place. Often these are groups that may not otherwise have been able to easily interact. These may include governmental organizations, research and development groups, component and material suppliers, assembly and testing equipment manufacturers, cell manufacturers, system integrators, and OEMs.

Modern day organizations such as the United States Advanced Battery Consortium LLC (USABC), the Electric Drive Transportation Association (EDTA), and the National Alliance for Advanced Battery Technology (NAATBatt) are actually successors to one of the earliest electric vehicle (EV) industry trade groups called the Electric Vehicle Association of America (EVAA) which was formed in 1909 and lasted for only seven years. EVAA members included electric power generating companies, vehicle manufacturers, and energy storage battery makers. Interestingly enough, if we compare the goals of some of these key organizations to the EVAA we find that there are some striking similarities (Table 1).

All three of these organizations have similar goals of working to bring together different elements of the battery market for the purpose of speeding the adoption of EVs in the market place. The other purpose, sometimes not clearly defined in all of them, is to encourage industry research and development activities in support of the goal of speeding up the technology-development rate in order to speed consumer adoption of these technologies.

Industry and government organizations also generate standards, which are important in emerging industries such as this because as has been said elsewhere "nature abhors a

The Handbook of Lithium-Ion Battery Pack Design. http://dx.doi.org/10.1016/B978-0-12-801456-1.00013-0

Table 1: Comparing industry trade groups

EVAA (1909–1917)	USABC (1991–today)	NAATBatt (2003–today)
To encourage the adoption and use of electric commercial and pleasure vehicle by electric light and power stations and their customers. (Kirsch, 2000)	… to promote long-term R&D within the domestic electrochemical energy storage (EES) industry and to maintain a consortium that engages automobile manufacturers, EES manufacturers, the National Laboratories, universities, and other key stakeholders. (USCAR, 2014)	… to grow the North American market for products incorporating advanced energy storage technology and to reduce the cost of those products to U.S. consumers. (NAATBatt, 2014)

vacuum." By this I mean that without some level of organization and direction, manufacturers will develop different products at all levels from cell to module to pack and system, but without any commonality between them. So without some level of guidance internal industry regulation and standard setting, governmental organizations are likely to step in and create rules and standards that may or may not be right for the industry. Therefore, it is often in the industry's best interest to self-regulate.

These include both national and international organizations that work collaboratively on developing and implementing standards around different areas of the battery, its performance, safety, recycling, and applications. Some organizations, such as the SAE, are focused on creating design, testing, and engineering standards to ensure that there is an amount of commonality among components such as on charging requirements, connectors, communication protocols, and other systems and components. Other organizations such as the United Nations (UN) and the Underwriters Laboratories (UL) are focused on testing and certification standards. Other organizations are focused on creating groups that enable networking between the different members of the value chain in order to find new and meaningful ways to help the industry grow.

One of the biggest challenges with standards organizations is the potential for overlap. In the event that multiple organizations generate a set of standards to cover the same area or product, but that also differ—which one would take precedent? For that reason, many of the standards organizations participate in each other's work groups and try to create unofficial "boundaries" for themselves to help prevent that overlap or even to coauthor standards.

This following section will review and discuss some of, but not all, the major organizations in each of these areas broken out into two main categories: voluntary standards, those that offer "best practices"; and mandatory standards, those that are required in order to build, sell, or use the product in certain markets.

Voluntary Standards

Society of Automotive Engineers (SAE)

The SAE is an international organization with more than 120,000 members that was founded in 1907 and is focused on three industry sectors: aerospace, automotive, and commercial vehicles (Society for Automotive Engineers, 2014). The SAE is one of the largest organizations of engineering professionals in the world which has two stated purposes. The first is to act as a professional organization for those working in the engineering fields and second is to act as a voluntary standards development organization. By involving a large number of engineering experts in their standardization efforts the SAE has been able to develop standards for virtually all aspects of automotive, aerospace, and commercial vehicles and already offer many standards surrounding vehicle electrification. Today there are about twenty different working groups within the battery standards committee working on developing and updating electrification standards.

One of the big benefits of the SAE standards is that they are quite affordable as they tend to run less than $100 each and with an SAE membership then can be significantly less than that. Additionally the standards are developed and updated by experts in the specific areas. So even as a small company, you can access the collective knowledge and best practices of a large group of industry technical experts and engineers.

With the rapidly growing interest in vehicle electrification, the SAE formed a battery standards development committee in 2009 that, even today, continues to work to develop and update standards that can be applied to automotive electrification in areas such as:

- Battery standards testing
- Advanced battery concepts
- Battery testing equipment
- Battery safety
- Battery packaging
- Small task-oriented vehicle batteries
- Hybrid battery technology
- Battery materials testing
- Battery labeling
- Battery terminology
- Battery electronic fuel gauging
- Battery transport
- Starter battery
- Battery recycling
- Truck and bus batteries
- Thermal management

These work groups, among others, have developed a significant set of standards. The following table is, at the time of this writing, a listing of the standards that SAE has either published or is currently working on targeted batteries for automotive electrification. There are also a large amount of additional standards that relate to hybrid and EVs but which are not battery focused, some of which are included in this list. It is also important to note that many of the SAE committees include participation from other industry groups such as ISO, International Electrotechnical Commission (IEC), UL, and others in order to ensure that there is no overlapping of standards and that any new standards will fit within a greater role in the industry.

SAE Standard	Standard Title	Objective
J1711	Recommended Practice for Measuring the Exhaust Emissions and Fuel Economy of Hybrid Electric Vehicles, Including Plug-in Hybrid Vehicles	Establishes uniform chassis dynamometer test procedures for hybrid electric vehicles (HEVs). The procedure provides instructions for measuring and calculating the exhaust emissions and fuel economy of HEV's.
J1715/2 (WIP)	Battery Terminology	Define common terminology for automotive electrochemical energy storage systems at all levels; component, subcomponent, subsystem, and system-level architectures including terms pertaining to testing, measurement, and system function related to energy storage.
J1797	Recommended Practice for Packaging of Electric Vehicle Battery Modules	Provides for common battery designs through the description of dimensions, termination, retention, venting system, and other features required in an electric vehicle (EV) application.
J2288	Life Cycle Testing of Electric Vehicle Battery Modules	Defines a standardized test method to determine the expected service life, in cycles, of EV battery modules.
J2289	Electric-Drive Battery Pack system: Functional Guidelines	Describes common practices for design of battery systems for vehicles that utilize a rechargeable battery to provide or recover all or some traction energy for an electric drive system.
J2293/1	Energy Transfer System for Electric Vehicles—Part 1: Functional Requirements and System Architectures	Establishes requirements for EV and the off-board Electric Vehicle Supply Equipment (EVSE) used to transfer electrical energy to an EV from an Electric Utility Power System (Utility).
J2293/2	Energy Transfer System for Electric Vehicles—Part 2: Communication Requirements and Network Architecture	Establishes requirements for EV and the off-board EVSE used to transfer electrical energy to an EV from an electric Utility Power System (Utility).
J2344	Guidelines for Electric Vehicle Safety	Identifies and defines the preferred technical guidelines relating to safety for EVs during normal operation and charging.
J2380	Vibration Testing of Electric Vehicle Batteries	Describes the vibration durability testing of a single battery (test unit) consisting of either an EV battery module or an EV battery pack.
J2464	Electric and Hybrid Electric Vehicle Rechargeable Energy Storage System (RESS) Safety and Abuse Testing	Describes a body of tests which may be used as needed for abuse testing of electric or hybrid EV batteries to determine the response of such batteries to conditions or events which are beyond their normal operating range.
J2908 (WIP)	Power Rating Method for Hybrid Electric and Battery Electric Vehicle Propulsion	Test method and conditions for rating performance of complete hybrid electric and battery EV propulsion systems reflecting thermal and battery capabilities and limitations.
J2910(WIP)	Design and Test of Hybrid Electric Trucks and Buses for Electrical safety	Covers the aspects of the design and test of class 4 through 8 hybrid electric trucks and buses.

—Cont'd

SAE Standard	Standard Title	Objective
J2936 (WIP)	Vehicle Battery Labeling Guidelines	Labeling guidelines for any electrical storage device at all levels of subcomponent, component, subsystem and system-level architectures describing content, placement, and durability requirements of labels.
J2946 (WIP)	Battery Electronic Fuel Gauging Recommended Practices	Recommend practice associated with reporting the vehicle's (hybrid and pure electric) battery pack performance details to the automobile user.
J2950 (WIP)	Recommended Practices (RP) for Transportation and Handling of Automotive-Type Rechargeable Energy Storage Systems (RESS)	Recommended Practices associated with identification, handling, and shipping of uninstalled RESSs to/from specified locations (types) required for the appropriate disposition of new and used items.
J2953 (WIP)	Plug-In Electric Vehicle (PEV) Interoperability with Electric Vehicle Supply Equipment (EVSE)	Establishes the interoperability requirements and specifications for the communication systems between PEV and EVSE for multiple suppliers.
J2954 (WIP)	Wireless Charging of Electric and Plug-In Hybrid Vehicles	Establishes minimum performance and safety criteria for wireless charging of electric and plug-in vehicles.
J2974 (WIP)	Technical Information Report on Automotive Battery Recycling	This SAE Technical Information Report provides information on Automotive Battery Recycling. This document provides a compilation of current recycling definitions, technologies and flow sheets, and their application to different battery chemistries.
J2983 (WIP)	Recommended Practice for Determining Material Properties of Li-Battery Separator	This SAE RP provides a set of test methods and practices for the characterization of key properties of Li-battery separator. It is not within the scope of this document to establish criteria for the test results, as this is usually established between the vendor and customer.
J2984 (WIP)	Identification of Transportation Battery Systems for Recycling Recommended Practice	The chemistry identification system is intended to support the proper and efficient recycling of rechargeable battery systems used in transportation applications with a maximum voltage greater than 12 V (including SLI batteries). Other battery systems such as nonrechargeable batteries, batteries contained in electronics, and telecom/utility batteries are not considered in the development of this specification. This does not preclude these systems from adapting the format proposed if they so choose.
J2990 (WIP)	Hybrid and EV First and Second Responder Recommended Practice	This RP describes the potential consequences associated with hazards from electrified vehicles and suggests common procedures to help protect emergency responders, recovery, tow, storage, repair, and salvage personnel after an incident has occurred with an electrified vehicle.

Continued

—Cont'd

SAE Standard	Standard Title	Objective
J2991 (WIP)	Range Test Protocol for PEV (Plug-In Electric Vehicles) Small Task Oriented Vehicles (STOV)	This test protocol is being developed to create a voluntary guideline for manufacturers of PEV STOV's to use to validate the range of their vehicles. The intent is to develop a laboratory test protocol for range testing that is repeatable and can be conducted using common dynamometer testing facilities.
J3097 (WIP)	Standards for Battery Secondary Use	To develop standards for a testing and identity regimen to define batteries for variable safe reuse. Utilize existing or in process standards such as Transportation, Labeling and State of Health. Add to these reference standards the required information to provide a safe and reliable usage.
J3004(WIP)	Standardization of Battery Packs for Electric and Hybrid Trucks and Busses	Identify existing standards and provide recommendations on design criteria and future standardization opportunities for battery packs on electric and hybrid truck and bus applications.
J3009 (WIP)	Stranded Energy—Reporting and Extraction from Vehicle Electrochemical Storage Systems	The intent of this document is to consider the type of information reported by the battery management system (BMS) and recommended discharge level dependent on a collision or vehicle fire. The document does not describe how the energy should be extracted.
J3012 (WIP)	Storage Batteries—Lithium-Ion Type	This document will focus on (1) product and functional definitions that describe the uniqueness and similarity of lithium-based technology versus conventional lead acid, and (2) development of new test procedures for performance and life cycle evaluation to establish new baseline for future nonconventional storage technology.

(SAE International, 2014).

By the time of this printing, this list will already be outdated so I strongly recommend that you check out SAE's Web site to find the latest and the greatest in the areas of your interest.

International Organization for Standardization

The ISO was formed in 1947 and is another global organization that is focused on creating voluntary industry standards. Where the SAE is focused on only automotive, aerospace, and commercial vehicle industries, ISO has a much wider focus including health, sustainable development, food, water, cars, climate change, energy efficiency and renewables, and services (International Organization for Standardization, 2014).

ISO standards tend to be a bit more expensive to purchase than SAE generally costing between several hundred to several thousand dollars to acquire a single standard. The ISO standards are also developed by a group of industry experts who work with a technical committee to gain consensus on the scope and content of the standard. Once the technical committee agrees on the standard, it passes to the larger ISO community for approval. Perhaps the biggest difference with SAE standards is that in order to be involved in the standard setting process your company needs to be a member of ISO.

The ISO often works jointly with the IEC and other standards organizations to develop some of their industry standards.

A partial list of EV battery standards follows:

- ISO 16750-1: Road vehicles—Environmental conditions for testing for electrical and electronic equipment—Part 1: General
- ISO 16750-2: Road vehicles—Environmental conditions for testing for electrical and electronic equipment—Part 2: Electrical Loads
- ISO 16750-3: Road vehicles—Environmental conditions for testing for electrical and electronic equipment—Part 3: Mechanical Loads
- ISO 16750-4: Road vehicles—Environmental conditions for testing for electrical and electronic equipment—Part 4: Climatic Loads
- ISO 16750-5: Road vehicles—Environmental conditions for testing for electrical and electronic equipment—Part 5: Chemical Loads

International Electrotechnical Commission

Formed in 1906, the IEC is a "sister" standard organization to ISO whose primary goal is to prepare and publish international standards for all electrical, electronic, and related technologies. The difference with the IEC's members is that they are not individual companies, but rather countries. Each member country has an equal vote in the standard development process. However, adoption of any standard by a member country is entirely voluntary. Member countries form national committees that are responsible for coordinating that country's interests in the standard setting process. These committees are typically made up of experts from government, universities, research institutions, and industry.

In the area of lithium-ion batteries, the IEC has developed standards including:

- IEC 60050-482—International Electrotechnical Vocabulary—Part 482: Primary and secondary cells and batteries
- IEC 61427-1—Secondary cells and batteries for renewable energy storage—General requirements and methods of test—Part 1: Photovoltaic off-grid application
- IEC 61429—Marking of secondary cells and batteries with the international recycling symbol ISO 7000-1135
- IEC 61959—Secondary cells and batteries containing alkaline or other nonacid electrolytes—Mechanical tests for sealed portable secondary cells and batteries

- IEC 61960–Secondary cells and batteries containing alkaline or other nonacid electrolytes—Secondary lithium cells and batteries for portable applications
- IEC 61982—Secondary batteries (except lithium) for the propulsion of electric road vehicles—Performance and endurance tests
- IEC 62133—Secondary cells and batteries containing alkaline or other nonacid electrolytes—Safety requirements for portable sealed secondary cells, and for batteries made from them, for use in portable applications
- IEC 62281—Safety of primary and secondary lithium cells and batteries during transport
- IEC 62485-2—Safety requirements for secondary batteries and battery installations—Part 2: Stationary batteries
- IEC 62485-3—Safety requirements for secondary batteries and battery installations—Part 3: Traction batteries

As with some of the other lists of standards, this is not all inclusive and by the time this book reaches print these may change and new standards may be added. So once again I encourage the reader to seek out the IEC Web site and investigate the relevant standards that are required in the countries where your products are intended to be manufactured and sold.

Institute of Electrical and Electronics Engineers

The Institute of Electrical and Electronics Engineers (IEEE) is a technical professional society that grew out of two earlier industry organizations. It brings together professionals involved in all aspects of the electrical, electronic, and computing fields and related areas (IEEE, 2014).

In essence, the IEEE is much like SAE; both are professional organizations and both create industry standards. The main difference is the focus, the IEEE covers many industries but only the electronics components and equipment.

The first thing that jumps out from a review of the battery-related standards is that almost all of the IEEE battery standards are focused on: a) stationary applications, b) lead acid battery, and c) nickel–cadmium chemistries. The following list is a partial list of the battery standards that the IEEE has created related to batteries.

- 450-2010—Recommended Practice for Maintenance, Testing, and Replacement of Vented Lead-Acid Batteries for Stationary Applications
- 484-2002—Recommended Practice for Installation Design and Installation of Vented Lead-Acid Batteries for Stationary Applications
- 485-2010—Recommended Practice for Sizing Lead-Acid Batteries for Stationary Applications
- 937-2007—Recommended Practice for Installation and Maintenance of Lead-Acid Batteries for Photovoltaic (PV) Systems

- 1013-2007—Recommended Practice for Sizing Lead-Acid Batteries for Stand-Alone Photovoltaic (PV) Systems
- 1106-2005—Recommended Practice for Installation, Maintenance, Testing, and Replacement of Vented Nickel-Cadmium Batteries for Stationary Applications
- 1115-2000—Recommended Practice for Sizing Nickel–Cadmium Batteries for Stationary Applications
- 1189-2007—Guide for Selection of Valve-Regulated Lead-Acid (VRLA) Batteries for Stationary Applications
- 1184-2006—Guide for Batteries for Uninterruptible Power Supply Systems
- 1187-2013—Recommended Practice for Installation Design and Installation of Valve-Regulated Lead-Acid Batteries for Stationary Applications
- 1361-2014—Guide for Selection, Charging, Test and Evaluation of Lead-Acid Batteries Used in Stand-Alone Photovoltaic (PV) Systems
- 1375-1998—Guide for the Protection of Stationary Battery Systems
- 1491-2012—Guide for Selection and Use of Battery Monitoring Equipment in Stationary Applications
- 1536-2002—Rail Transit Vehicle Battery Physical Interface
- 1561-2007—Guide for Optimizing the Performance and Life of Lead-Acid Batteries in Remote Hybrid Power Systems
- 1562-2007—Guide for Array and Battery Sizing in Stand-Alone Photovoltaic (PV) Systems
- 1568-2003—Recommended Practice for Electrical Sizing of Nickel–Cadmium Batteries for Rail Passenger Vehicles
- 1578-2007—Recommended Practice for Stationary Battery Electrolyte Spill Containment and Management
- 1625-2008—Standard for Rechargeable Batteries for Portable Computing
- 1635-2012—Guide for the Ventilation and Thermal Management of Batteries for Stationary Applications
- 1657-2009—Recommended Practice for Personnel Qualifications for Installation and Maintenance of Stationary Batteries
- 1660-2008—Guide for Application and Management of Stationary Batteries Used in Cycling Service
- 1661-2007—Guide for Test and Evaluation of Lead-Acid Batteries Used in Photovoltaic (PV) Hybrid Power Systems
- 1679-2010—Recommended Practice for the Characterization and Evaluation of Emerging Energy Storage Technologies in Stationary Applications
- 1725-2011—Standard for Rechargeable Batteries for Cellular Telephones
- 1825—Standard for Rechargeable Batteries for Digital Cameras and Camcorders
- 2030-2011—Guide for Smart Grid Interoperability of Energy Technology and Information Technology Operation with the Electric Power System (EPS), End Use Applications, and Loads

As shown in this list the IEEE does have several standards for lithium-ion batteries, but they are targeted at portable power applications such as cellular telephones and portable computer batteries (IEEE, 2014).

Underwriter's Laboratory

The UL is one of the oldest private safety testing and certification organizations in the world. Founded in 1894, the UL has focused their efforts on safety testing and certification for consumer products. Today UL continues to increase their testing scope to include significant consulting business in order to "…keep safety ahead of innovation in an evolving global landscape" (Underwriter's Laboratory, 2014).

UL testing and certification has been well established for consumer batteries such as the battery in your laptop, tablet, or smart phone. However, the automotive industry has initially pushed back on the need for UL testing and certification for automotive batteries. The reasons for this are twofold. First, the automotive manufacturers already do significant and expensive testing to validate their batteries and have developed their own set of testing and verification-requirements that are aligned with the National Highway Transportation Safety Agency (NHTSA) requirements. Second, they did not like the idea of a separate organization that had little involvement in the product development to have the final approval. And of course UL testing adds significant expense to an already expensive automotive validation program which would require paying someone for testing that may have already been completed. UL testing must be conducted by the Underwriter's Laboratory, whereas other certifications such as UN testing may be self-certified requiring only that you have the data from the testing available.

As of this writing there are seven UL certification requirements for batteries, including:

- UL 1642 Lithium Cell
- UL 1973 Batteries for use in Light Electric Rail applications and Stationary Applications
- UL 1989 Standby Batteries
- UL 2054 Alkaline Cell or Lithium/Alkaline Packs
- UL 2271 Batteries for use in Light Electric Vehicle (LEV) applications
- UL 2580 Batteries for use in Electric Vehicles
- UL/CSA/IEC 60065 Batteries used in audio and video equipment (Underwriter's Laboratory, 2014)

As is shown in the list above, the range of UL certifications covers diverse areas such as lithium cells, rail, light vehicles, stationary energy storage, standby batteries, and EVs.

Det Norske Veritas (DNV-GL)

Founded in 1864 in Oslo, Norway, Det Norske Veritas (DNV) is a key organization focused on providing classification, technical assurance, and independent expert advisory services to the maritime, oil and gas, and energy industries. In late 2013, DNV-GL developed and published a draft guideline for large batteries for maritime applications titled "*DNV GL Guideline for Large Maritime Battery Systems*" (DNV GL, 2013). The purpose of the DNV-GL guideline is "...to help ship owners, designers, yards, system- and battery vendors and third parties in the process of feasibility study, outline specification, design, procurement, fabrication, installation, operation, and maintenance of large lithium-ion based battery systems" (DNV GL, 2013).

With the increase in electrification of maritime and shipping applications and increased regulatory rules aimed at reducing pollution around the world's major ports, many ship builders are looking to replace diesels with battery power. With the strong relationship and long history that DNV-GL has with this industry, their guideline has set the standard for these applications. In addition, DNV offers third-party testing and certification services.

Research and Development and Trade Groups

USABC/USCAR

The USABC is a subgroup of United States Council for Automotive Research LLC (USCAR). USCAR was formed in 1992 with the goal of strengthening the technology base of the U.S. auto industry through cooperative research and development with participation from General Motors, Ford, and Chrysler. The USABC was actually formed in 1991 a year earlier than USCAR with membership representing General Motors, Ford, and Chrysler with the goal of promoting "...long-term R&D within the domestic electrochemical energy storage (EES) industry and to maintain a consortium that engages automobile manufacturers, EES manufacturers, the National Laboratories, universities, and other key stakeholders" (USCAR, 2014).

USABC has created several set of performance requirements for 12V (Appendix A), 48V (Appendix B), HEV (Appendix C), PHEV (Appendix D), and EV (Appendix E) automotive energy storage systems. Additionally, as part of their cooperative research work they frequently issue Request for Proposal Ideas which, in partnership with the U.S. government, provide funding for the development of advanced automotive battery systems with specific performance targets. Additionally, the USABC has developed several tools including a battery cost model, a battery calendar life estimator manual, and battery test manuals for both low voltage and high voltage systems.

NAATBatt

The National Association for Advanced Battery Technology (NAATBatt) is an industry trade group whose core mission is:

> *... to grow the North American market for products incorporating advanced energy storage technology and to reduce the cost of those products to U.S. consumers.*
>
> ***(NAATBatt, 2014)***

The NAATBatt organization was originally formed as a U.S. based trade group but today the membership has grown and now includes member organizations from countries including Canada, China, and Europe. The organization works to make connections between companies within the battery industry to help speed the introduction of energy storage systems into all types of products. More recently the organization has also begun to include other types of electrochemical storage technologies such as ultra-capacitors and super capacitors.

Portable Rechargeable Battery Association

Formed in 1991, the Portable Rechargeable Battery Association (PRBA) was founded by five of the major rechargeable battery manufacturers of that period including Energizer, Panasonic, SAFT America, SANYO Energy Corporation, and Varta Batteries. The PRBA was established as a nonprofit trade association to meet the growing need for industry-wide battery recycling programs.

The PRBA also acts as the voice of the rechargeable power industry, representing its members on legislative, regulatory, and standards issues at the U.S. state, federal and international level (Portable Rechargeable Battery Association, 2014). To date the PRBA has focused mainly on smaller battery powered applications and not on the larger automotive, industrial, or grid/stationary applications.

Light Electric Vehicle Association

The Light Electric Vehicle Association (LEVA) is an international trade association for LEVs. LEVA defines a light electric vehicle as an electric bicycle, scooter, motorcycle, three wheel vehicle, or light four wheel vehicle. The LEVA group is based in the U.S. but has membership in both Europe and Asia. With the very large usage of electric bicycles in Asia and Europe, this group is focused on supporting and growing this market by creating a group of retailers, dealers, distributors, manufacturers, and suppliers promoting the development, sale, and use of LEVs worldwide.

US National Labs

There are also several major U.S. national laboratories that have significant focus on battery research and testing, including Sandia National Lab (SNL), Oak Ridge National Labs (ORNL), the National Renewable Energy Center (NREL) in Golden, Colorado, the Pacific Northwest National Laboratory (PNNL), and Idaho National Laboratory (INL) all are major partners in the testing of advanced batteries. The National Labs do not do certification, but rather conduct various testing, competitive assessments and evaluations and research partnering with industry, academia, and other government organizations.

Mandatory Standards Organizations

United Nations (UN) 38.3

Perhaps the single most vital safety testing standard is the UN "*Recommendations on the Transport of Dangerous Goods: Manual of Tests and Criteria.*" as mentioned in Chapter 12 section 38.3 of this document, defines the testing standard that a lithium metal or lithium-ion battery cell, module, or pack must pass prior to being approved for shipping. The document includes the testing methodology, the testing quantities, and identifies the passing criteria. For lithium-ion batteries, there are a set of eight (8) tests that all lithium-ion or lithium metal batteries must pass prior to being approved for shipping via any method. These tests include:

- Test T.1: Altitude simulation
- Test T.2: Thermal test
- Test T.3: Vibration
- Test T.4: Shock
- Test T.5: External short circuit
- Test T.6: Impact/crash
- Test T.7: Overcharge
- Test T.8: Forced discharge

It is necessary to understand whether you are testing a cell, a module, or a pack and whether it is rechargeable or not as this determines how many of the tests that must be conducted and must be passed. All lithium-ion and lithium metal cells must pass T1 through T6 and T8. The overcharge test is excluded for cells as there would be no way to provide a charge to an unassembled cell during transport. All nonrechargeable cells must pass T1 through T5. All rechargeable battery types, which for this purpose means any assembly of more than one cell including bricks, blocks, modules, and packs, are required to be tested to T1 through T5 and T7 (United Nations, 2013).

This last definition is an important one, as the early publications of this requirement were not clear on the definitions and differences between cells and modules and pack. The new regulation states:

> *Units that are commonly referred to as "battery packs", "modules" or "battery assemblies" having the primary function of providing a source of power to another piece of equipment are for the purposes of the Model Regulations and this Manual treated as batteries.*
>
> ***(United Nations, 2013)***

The requirement also allows for the shipment of up to 100 prototypes via truck per year. This was added as the original regulation created a bit of a chicken and egg scenario, as the original regulation did not allow for any shipments without certification, but that also eliminated the ability for manufacturers to actually get product to the testing houses.

Chinese Standards and Industry Organizations

The Standardization Administration of China (SAC) is responsible for managing, supervising, and coordinating standardization work in China. SAC was established in 2001 and authorized by the State Council to exercise administrative responsibilities by undertaking unified management, supervision and overall coordination of standardization work in China (Standardization Administration of the People's Republic of China, 2014).

SAC also represents China in several of the international standards organizations such as the ISO, the IEC and other international, and regional standardization organizations. SAC is responsible for approving and organizing international cooperation and exchanging projects on standardization (Standardization Administration of the People's Republic of China, 2014).

In China, one of the largest organizations that is working with both the vehicle and the battery manufacturers is the China Automotive Technology and Research Center (CATARC). Formed in 1985, CATARC works with the Chinese National government in the area of creating automotive standards and technical regulations, conducting product and quality system certification, conducting industry planning and policy research, as well as providing basic technology research in a multitude of areas. While continually working to support the Chinese regulations, CATARC also strives to maintain its position as an impartial, third-party standards, testing and certification organization (China Automotive Technology and Research Center, n.d.).

The National Technical Committee on Automotive Standardization (NTCAS) is responsible for all automotive standards including subcommittee 27 which is responsible specifically for EV standardization. The NTCAS also works with ISO, and the secretariat of NTCAS is located at the CATARC headquarters in Tianjin, China.

China's GB standards correspond to the major ISO and SAE standards. Table 2 below shows the corresponding Chinese and international standards related to all electrification components.

In respect to Chinese testing and certification organizations, the two most highly accepted organizations are CATARC and the Battery Testing Lab within the North Vehicle Research Institute 201. Both can carry out battery testing (mainly according to QC/T 743) and issue official certifications.

European Standards and Industry Organizations

International Battery and Energy Storage Alliance

Formed in 2013 the International Battery and Energy Storage Alliance (IBESA) is a European trade group which was formed by the International Photovoltaic Equipment Association (IPVEA) and solar research group EuPD as the first international alliance for providers of battery

Table 2: Chinese EV standards

	Standard No	Standard Name	Reference
1	GB/T 18384.1-2001	Electric vehicles—Safety specification Part 1: Onboard energy storage	ISO/DIS 6469.1:2000
2	GB/T 18384.2-2001	Electric vehicles—Safety specification Part 2: Functional safety means and protection against failures	ISO/DIS 6469.2:2000
3	GB/T 18384.3-2002	Electric vehicles—Safety specification Part 3: Protection of persons against electric hazards	ISO/DIS 6469.3:2000
4	GB/T 4094.2-2005	Electric vehicles—Symbols for controls, indicators, and telltales	ISO 2575:2000/ Amd.4:2001; JEVS Z 804:1998
5	QC/T 743-2006	Lithium-ion Storage Battery for Electric Automotive	N/A
6	GB/T 18385-2005	Electric vehicles Power performance Test method	ISO/DIS8715:2001
7	GB/T 18386-2005	Electric vehicles Energy consumption and range Test procedures	ISO 8714:2002
8	GB/T 18387-2008	Limits and test methods of magnetic and electric field strength from electric vehicles, Broadband, 9 kHz to 30 MHz	SAE J551/5 JAN 2004
9	GB/T 18388-2005	Electric vehicles-Engineering approval evaluation program	N/A
10	GB/T 18488.1-2006	The electrical machines and controllers for electric vehicles—Part 1—General specification	IEC 60785:1984 IEC 60786:1984 IEC 60034-1:1996
11	GB/T 18488.2-2006	The electrical machines and controllers for electric vehicles—Part 2: Test methods	JEVS E701-1994
12	GB/T 19836-2005	Instrumentation for electric vehicles	IEC 784:198
13	GB/T 18487.1-2001	Electric vehicle conductive charging system—Part 1: General requirements	IEC 61851-1:2001
14	GB/T 18487.2-2001	Electric vehicle conductive charging system—Electric vehicles requirements for conductive connection to an A.C./D.C. Supply	IEC/CDV 61851-2-1:1999
15	GB/T 18487.3-2001	Electric vehicle conductive charging system—A.C./D.C. electric vehicle charging station	IEC/CDV 61851-2-2:1999 IEC/CDV 61851-2-3:1999 JEVS G101-1993 SAE-J 1772-1996
16	GB/T 20234.1-2011	Connection set of conductive charging for electric vehicles—Part 1: General requirements	IEC 62196
17	GB/T 20234.2-2011	Connection set of conductive charging for electric vehicles—Part 2:AC charging coupler	IEC 62196
18	GB/T 20234.3-2011	Connection set of conductive charging for electric vehicles—Part 3:DC charging coupler	IEC 62196
19	GB/T 24552-2009	Electric vehicles—Windshield demisters and defrosters system—Performance requirements and test methods	N/A
20	GB/T 19596-2004	Terminology of electric vehicles	SAE J1715/2

and electrical energy storage solutions to represent the interests of the emerging solar storage sector in respect to battery and energy storage requirements. The vision of the IBESA is to:

> *...promote a path of cooperation and mutual support in achieving proactive solutions between all sectors within the Photovoltaic (PV) Power Generation, Battery Storage and the Smart Grid Technology value chain.*
>
> **(CHoehner Research & Consulting Group GmbH, 2013)**

With the significant amount of PV generation that is in place in Europe, the addition of energy storage was a natural growth area. The IBESA brings together members of the industry from all areas of the value chain to help share knowledge among its more than 70 member companies.

EUROBAT

Within the European Union, another organization that is working hard to bring together members of the battery community is EUROBAT. EUROBAT is the trade association of European automotive, stationary, industrial, and battery manufacturers. It is an industry trade organization made up of over 40 members from various parts of the battery and automotive industry from across Europe. The goal of the group is to act as a single voice to promote the interests of European manufacturers (EUROBAT, 2010). EUROBAT represents lead-acid, nickel, lithium and sodium based chemistries within their organization.

Verband der Automobilindustrie

Within Germany, the Verband der Automobilindustrie (VDA) is an automotive industry trade group that dates back over 100 years with its current organization structure and name dating back to 1946. The name translates to the German Association of the Automotive Industry which is made up of over 600 member companies with the goal of conducting research and production of clean and safe automotive transportation battery solutions. In addition to these activities, the VDA is the annual sponsor for the International Auto Show held in Frankfurt every year. In regard to their work around batteries, the VDA has been working with ISO to create a set of lithium-ion cell size standards (Verband der Automobilindustrie, 2014).

Design Guidelines and Best Practices

- There are a wide variety of voluntary standards organizations that must be evaluated in designing a new energy storage system including: SAE, ISO, IEC, IEEE, UL, and DNV-GL
- Trade groups and research organizations include USABC/USCAR, NAATBatt, PRBA, and LEVA
- U.S. National labs doing work on energy storage systems include: SNL, ORNL, NREL in Golden, Colorado, PNNL, and INL
- China standards and certification organizations include: SAC, CATARC
- European battery standard organizations include: IBESA, EUROBAT, and VDA

CHAPTER 14

Second Life and Recycling of Lithium-Ion Batteries

In this chapter we will review the concepts of lithium-ion battery recycling, reuse, repair, remanufacturing, and second life applications. This topic is still very much in its early stages of development as there are currently no recycling requirements or regulations for lithium-ion batteries such as those that are in place for lead acid batteries. The development of second life applications of used lithium-ion batteries is also still in its very early days. Because each battery pack is unique and with different modules in every pack, there is a major challenge in trying to mix and match different battery modules and chemistries into second life applications. In addition to this, if the prices of first life lithium-ion batteries drop to the levels such as the U.S. DOE and USABC are targeting ($100/kWh and $150/kWh, respectively) it may make second life batteries noncompetitive from a price perspective. However, with current pricing trends it may be a while before this happens, thereby making second life batteries viable in the near term.

In addition to lack of recycling standards, the supply of used lithium-ion batteries that are coming into the market is still very small. As most of the vehicle applications that are using lithium-ion batteries have only been recently introduced into the market place the batteries in them are still in the early phase of their life. That means that it will likely be another five years or more before these batteries reach their end of life and are ready to enter into the post-vehicle value stream for either recycling or being rebuilt for second life applications. And while the packs are still in early stages of commercialization, there are still some packs that are beginning to enter the value stream from testing, certification, and validation from the pack and vehicle manufacturer testing.

However, while lithium-ion batteries are still only just beginning their first life, there are plenty of nickel metal hydride batteries that are beginning to come to the end of their useful life. This may enable the industry to put a recycling and second life infrastructure in place and allow it to begin to grow in preparation for the coming of the lithium-ion batteries.

The value proposition for both of these applications is still not extremely strong, especially as it relates to second life batteries. However, it is feasible to imagine a future where the cost of the lithium-ion battery pack in your vehicle includes a "core charge" not unlike the lead acid battery. This essentially represents the value of the battery at the end of its first life that the consumer could expect to get back. Think of this in the same way you do the lead acid starter battery in your car. This could present some interesting business models to the original

The Handbook of Lithium-Ion Battery Pack Design. http://dx.doi.org/10.1016/B978-0-12-801456-1.00014-2

equipment manufacturers (OEMs) and dealerships in the future as well. For instance, the OEM could carry the "core charge" thereby reducing the price to the consumer with an agreement of some sort that the battery must come back to one of the OEM dealers once it reaches its end of life.

The key to being successful, especially in this early stage of the industry, is being able to perform multiple processes on the incoming battery packs. In fact the mantra of the industry is summed up by some experts as being related to the "four R's": repairing, remanufacturing, refurbishing and repurposing used battery packs. In fact we may add a fifth "R" to that list and include recycling to round out the value chain.

One of the challenges that the industry may still face is in respect to the legal aspects of these processes. It will be important to define and answer the question "who is responsible?" when a remanufactured or rebuilt battery pack fails. Is it the original battery manufacturer that holds the warranty and does the original warranty transfer to the remanufacturer? The second use customer? Once the battery is tested and remanufactured or rebuilt who will warranty it now? The cell manufacturer? In order for this industry to develop and evolve, it will be necessary for these legal hurdles to be understood and to have answers in place in order for the industry to grow.

Repairing and Remanufacturing

Perhaps one of the most undervalued yet most important aspects of the used battery process is being able to repair and remanufacture battery packs. This could actually become a vital aspect of the service industry. While large automotive manufacturers may have infrastructure to manage their own service parts and warranty, smaller companies and battery manufacturers will not have the infrastructure to be able to manage these functions. Therefore, we may see the growth of a service industry that is capable of acting as the service, repair, and warranty support for the battery industry players. The other benefit of this is that it will consolidate the capital investment as well as the expertise into a few companies.

One of the challenges with repair and service parts is maintaining an inventory. One benefit of moving to an industry service model is that it becomes more feasible to envision a model that allows for batteries to receive regular charging service in this consolidated industry model. Since lithium-ion batteries tend to lose capacity by up to a couple of percent per month it becomes necessary to regularly charge them in order to avoid that capacity loss becoming permanent. This may be easier to accomplish in a facility dedicated to maintaining these batteries.

Refurbishing, Repurposing, and Second Life

With all this being said there could be some very promising opportunities to use "second life" batteries and to use recycled materials to reduce costs. Remember that most battery pack manufacturing applications design their end of life to be achieved when they reach 80% of the original capacity or 80% of the original power.

So let's take a look at what that means. For example if we use a Nissan Leaf battery pack that has an original total capacity of 24 kWh, it will reach its end of life when it reaches ~19.2 kWh of total energy. That means that it will still have about ~15 kWh of usable energy, again assuming 80% depth of discharge. If the modules that are being used in that battery are removed and are refurbished, which we will talk about shortly, they can be re-installed into another unit and can continue to provide useful energy for low C-rate applications for years to come.

The challenge that we face today is that since there are no cell standards, every cell manufacturer has a different cell type and all of those different cell types require different module types and there are still a wide variety of cell chemistries being used out there. All of these things mean that it is very difficult to mix and match battery modules into second use applications. Many pack manufacturers also weld the cells into the module which means that it is nearly impossible to replace cells in a welded module, so the whole module must either be used or recycled.

The second life process and value chain for battery begins with the battery reaching 80% of its original power or capacity. As shown in Figure 1 below, the battery can then be removed

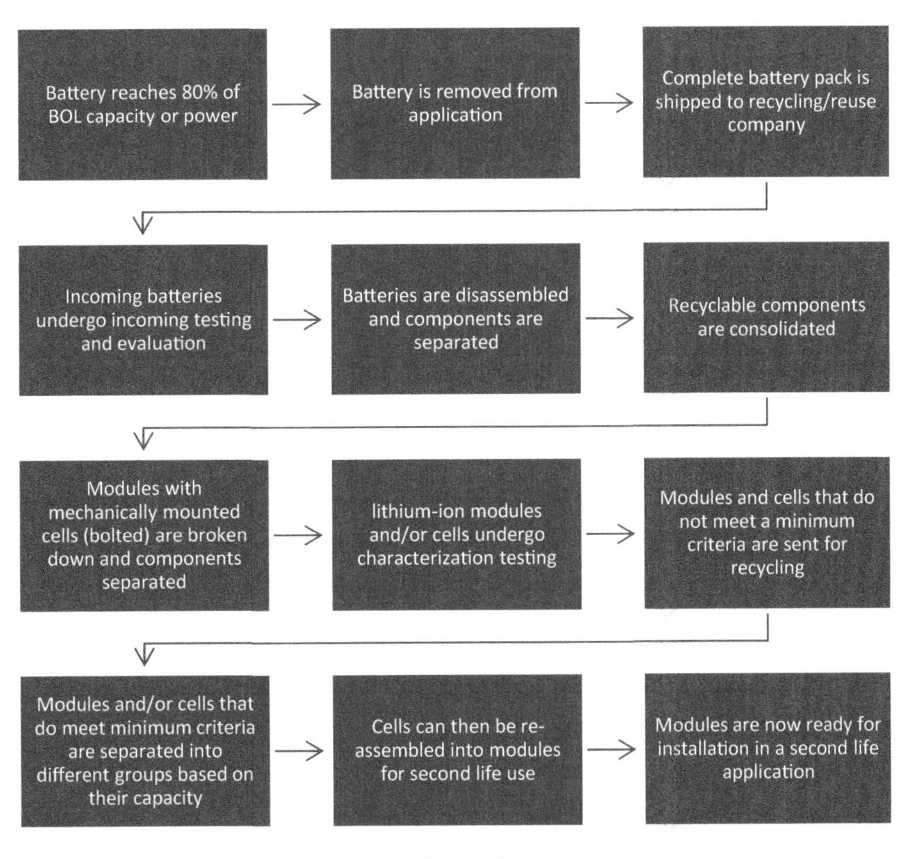

Figure 1
Battery second use life cycle. BOL, begin of life.

from the vehicle by a certified service center and shipped to a recycling and reuse company. Once there the batteries are evaluated and undergo initial testing to determine their capacity, voltage, and operating performance level. At this point the batteries can begin the recycling process by being disassembled. The various components are then separated into the different recycling value streams (plastics, metals, etc.). At this point modules that are built with cells that have been mechanically installed with bolts can be disassembled and the cells can begin undergoing characterization testing; modules that are built by welding the cells together will also undergo characterization testing. Any cells or modules that do not meet minimum performance criteria are then sent to be recycled, while those that do meet the criteria are then grouped within different capacity ranges and are ready for reinstallation and to begin their second life in a new application.

Perhaps one of the most important things to keep in mind when it comes to repurposing and reusing lithium-ion battery cells or modules is the need to characterize them. As was previously mentioned, when building battery packs the overall power and capacity will be limited by the lowest capacity cell in the pack. Similarly, when re-assembling packs with previously used cells or modules, the final pack will be limited by the lowest capacity cell, therefore the cells or modules must be closely matched in capacity and voltage. The other aspect of this is matching cells or modules based on some aging criteria. In the same way that a lower capacity cell will lower the final capacity of a pack, an "older" or more aged cell or module will limit the life of the newly built pack. So when evaluating cells and modules for second life applications, it is vital to ensure that they are matched not only on the basis of capacity but also based on their state of health (SOH), or in other words how close they are to their end of life.

This also applies to servicing of battery packs; you do not want to install a cell or module that is already at 70% of its original SOH into a pack that is near the beginning of its SOH. For instance, if you installed a module that was at 70% of its SOH into a battery pack that was near 90% of its SOH, the aged battery module would likely fail first thereby damaging the remaining modules or at the very least reducing their SOH. So when building or rebuilding modules you want to very closely match both the current capacity as well as the SOH of the cells and modules.

Perhaps the two biggest challenges that second life batteries face today are the supply of batteries for refurbishment and the cost of refurbishing them. With electric vehicles still making only a small portion of the current vehicles that are on the road and only a very small percentage of the annual sales, there simply are not enough vehicles on the road to generate an ample supply of batteries that *can* be refurbished. And since the actual life of these batteries is still somewhat of a mystery as the first of these vehicles that were sold into the market have not yet reached the end of their estimated life, we really will not know until they get there how many will begin coming into the recycling market. As I sit here writing this, it is 2014 only four years after the introduction of the Chevrolet Volt and the Nissan Leaf, which means that it will likely be another four to six years before those first vehicles reach the end of their warranty period. And of course that assumes that the batteries begin to fail about the same time as they end their warranties. If we compare this introduction period to the current average age of cars on the roads in the US, which is about 11 years today, it will not likely be until 2020 or 2021 that many of these batteries begin to enter the second life value chain. In addition, the volumes of the batteries available will be limited to the number of vehicles that were sold, which means that we are talking about initial numbers in the range of 20,000 to 30,000 battery packs beginning to feed their way in to the second life value stream in about 2018 and likely peaking around 2020. Now of course as new models continue to be introduced those volumes will continue to rise steadily. But for now, the second life use of automotive batteries will continue to be somewhat limited.

The second issue I mentioned is with the cost of refurbishing these used batteries. The end cost of the refurbished batteries needs to take into account the transportation, testing, remanufacturing, and reinstallation into the new application as well as the operating overhead, warranty, and of course a profit margin. Today with so few batteries on the market, the overhead costs are relatively high and must be amortized over a much smaller volume of products, which means that the initial prices will be high. They will be lower cost than new, but are still likely to be cost challenged until the volume of batteries on the market increases.

Second Life Partnerships

Several of the automakers have already created partnerships with companies that will convert their used batteries, many from initial test and demonstration fleets, into second life applications. Both General Motors and Nissan have announced partnerships to do just that. In 2012 General Motors and ABB announced a partnership to repackage Chevrolet Volt batteries into a new modular unit that would be capable of offering about 2 h of power. They are working on applying this technology to a wide range of stationary and grid based energy storage applications. This creates an initial second life "path" for the used Chevy Volt batteries (General Motors, 2012).

Similarly, Nissan has entered into a joint venture with the Sumitomo Corporation of Japan in order to develop large scale energy storage systems based on used Nissan Leaf batteries. The joint venture, named the 4R Energy Corporation, intends to demonstrate the use of large scale grid energy storage solutions to perform power smoothing and renewable solar power integration into the grid (Nissan, 2014).

Several universities are also working to install demonstration programs for not only first life energy storage systems but also second life systems in order to validate their use and value. One such program is housed at the University of California, San Diego. In addition to one of the largest installed fuel cell programs, a 30 kW solar storage array, flow battery installation, and a newly installed lithium-ion battery installations UCSD has installed a second life battery program. The program is funded by the Department of Energy and the National Renewable Energy Laboratory for endurance testing of second life electric vehicle batteries in stationary applications (UC San Diego, 2014).

Today there are only a couple of companies that are actively working on the "4R's" including Spiers New Technologies a former division of ATC New Technologies. The group was created with the specific focus to refurbishing and repairing of high voltage battery systems and modules; returning logistics of high voltage battery packs or modules from dealers and dismantlers; providing cell grading analysis and balancing of cells for remanufacturing and second life use; creating a second life market for graded cells; and recycling preparation (ATC New Technologies, 2014; Spiers, 2014).

Another company working similarly on the "4R's" for lithium-ion batteries is Sybesma's Electronics. Sybesma's grew out of an automotive and electronics repair company and now also offers lithium-ion and nickel-based battery refurbishment, repair, and recycling. Services offered here include returning logistics of high voltage battery modules from dealers and dismantlers, recycling preparation, cell grading and balancing of cells for secondary use, and creating a second life market (Sybesma's Electronics, 2014).

These two companies are at the forefront of the creation of a new second life, repair, refurbishment, and recycling industry in the US. If governmental regulations begin to get put in place to require recycling and second life for lithium-based battery companies such as these are in a very good position to lead the industry.

Recycling

Lithium-ion batteries and battery packs contain many materials that can be recycled, including lithium, cobalt, manganese, nickel, aluminum, copper, steel, and plastics. However, as it stands today there are no incentives nor regulations that would drive companies to recycle batteries and the financial gains still outweigh the benefits. Many battery manufacturers, both pack and cell, have taken it upon themselves to put agreements in place with companies to manage their waste stream including recycling battery components and in some cases complete packs.

One of the challenges with recycling lithium-ion battery cells is that it is very energy intensive to get the precious metals out of the cells and often uses more energy that can be recovered during the recycling process. In addition the value of the materials that are recovered currently does not cover the cost of the recycling process. However, there are several methods for effectively recycling lithium-ion batteries; they either use high temperatures or low temperatures to extract the precious materials. Using high temperature, cells are essentially melted so that the raw materials can be separated. This smelting process enables the cobalt, nickel, and copper to be recovered during the smelting process while lithium, aluminum, tin, and manganese are expelled in the slag which, while still expensive and energy intensive, can continue to be refined (Figure 2). Some of these materials can be refined for use in future batteries, however much of it is no longer of high enough quality for using in battery applications but can still find uses in many other applications.

In low temperature recycling, cells are essentially frozen using a cryogenic process and then shredded. From this point the materials can be separated using a series of filters, shakers, and other separating technologies. Compared to the high temperature processing, this technique is much less energy intensive. However, the quality of the materials must, again, be evaluated to determine if they can be reused in future battery applications, or if their quality is not high enough they may be used in other products or simply recycled.

And of course the third method for recovering materials from spent batteries is the physical separation process. This usually requires that the inputs be separated and a single type of battery chemistry (LFP, LCO, LMO, NMC, etc.) be introduced into the process. This may be done by hand and can be done at very low volumes but could enable electrolyte, anode material and potentially even cathode material to reenter the battery value chain while the remaining materials are recycled.

One challenge of battery recycling is the importance of being able to differentiate the chemistries and the materials. There have been numerous reports of lithium-ion batteries that have been developed in lead acid form factors being introduced into the lead acid recycling process. This does not end well and generally ends with some form of combustion as the two types of products are mixed. This is why it is important even at this early stage of the industry to begin developing standards for labeling lithium-ion batteries. The Society of Automotive Engineers has begun working on standards for labeling of lithium-ion batteries as well as for recycling them. As the number of these products that are entering the recycling process increases this will continue to gain importance in order to ensure the safety and security of the recycling industry and those people involved with the handling of these products.

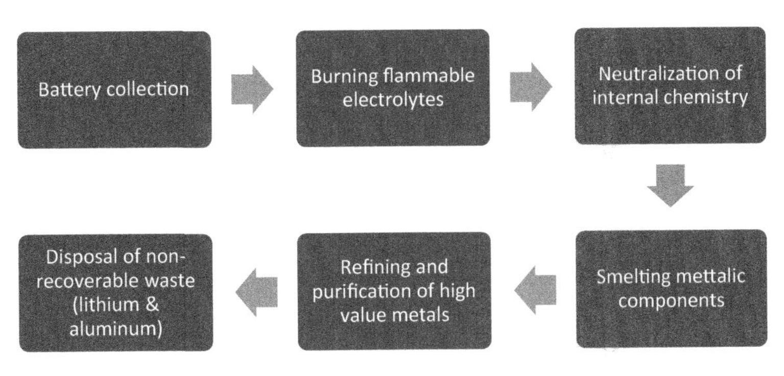

Figure 2
Lithium-ion battery recycling process.

Of course the other topic that we should discuss in relation to recycling of lithium-ion batteries is finding the value in the process. The value of the materials that are recovered during the recycling process is directly related to the value of the materials that go into it. So lithium-ion batteries with higher cost materials such as cobalt and nickel will offer more recoverable value. However lithium-ion batteries with iron phosphate or manganese that are already very low cost materials, will expend higher costs to recover very low cost materials—thus creating a negative cost proposition. Only once the value of the materials that are recovered is greater than the cost of recovering it and at the same time lower than the cost of "new" materials will recycling become a cost-effective solution.

Today there are only a couple companies that have begun to invest heavily in the recycling of lithium-ion batteries. Umicore is a German-based global materials company that is leading the market in lithium-ion battery recycling. They have invested heavily in a new plant in Hoboken, Germany in preparation and expectation of the growth of the market (Umicore, 2014).

In the US, the Department of Energy awarded a $9.5 million grant to a California-based company called Toxco to expand their facility in Ohio that was already recycling both lead acid and nickel-metal hydride automotive batteries. Toxco also happens to be one of the leading recyclers in North America and has been recycling both primary and secondary batteries from portable power and personal electronics devices in their facility in British Columbia, Canada since 1992 (Abuelsamid, 2009).

Right now these two companies are leading the market in terms of recycling for lithium-ion batteries and are likely to be well positioned to meet the future of battery recycling.

Design Guidelines and Best Practices

- The four "R's" of lithium-ion batteries are repairing, remanufacturing, refurbishing and re-purposing and recycling is the fifth "R"
- There are still few regulations related to the five "R's" in place today, the industry is still in its infancy
- The recycling process can be done either by high temperature or low temperature separation techniques
- Partnerships for second life of lithium-ion batteries are in the early stages of formation

CHAPTER 15

Lithium-Ion Battery Applications

Now that we have a better understanding of the parts of a lithium-ion battery, the basic design guidelines, and the testing needs, we will take a deeper look at how the battery is applied and used in the marketplace. This chapter will review lithium-ion battery-based applications from some of the smaller applications such as e-bikes and scooters to automotive microhybrids and electric vehicles (EVs), followed by industrial and commercial transportation, and then end with a review of some large energy storage systems ranging from community and household energy storage units to multi-megawatt-sized grid support systems. We will review these applications from the perspective of the battery not necessarily the end product (e.g., the bike, the car, the bus, the grid themselves will only be discussed at high level).

We will not cover "portable power"-type lithium-ion applications such as laptop computers, tablets, handheld games, power tools, smart phones or cell phones as these types of applications are covered in greater detail in other books and most will only require a couple of cells, a laptop may contain up to nine 18650-type cells for example. But keep in mind that all of the concepts discussed here so far will also apply to these types of applications, generally at a greatly scaled down level. Instead we will cover applications such as personal transportation, automotive, industrial applications, heavy-duty truck and bus, marine and maritime applications, grid and stationary applications, and aerospace applications of lithium-ion batteries those applications that will require very large collections of lithium-ion cells. This will not be an all-inclusive list but will attempt to give the reader a good understanding of the wide variety of applications that are using lithium-ion batteries.

Personal Transportation Applications

Personal transportation is the largest market for electrified transportation in the world. The largest portion of this market is made up of electric bicycles, scooters, and motorcycles. In many areas of the world, but largely in the Asian countries, electric bicycles are in common use as daily forms of transportation with more than 30 million electric bicycles being sold annually. In Europe and the United States, most of the electric bikes that are in use are used for sporting enthusiasts rather than for daily transportation but those markets are also beginning to experience significant growth.

The first category of electric bicycles are those called "pedelecs." A pedelec includes a small electric motor and battery that provides the rider with power assistance during riding. The pedelec is classified as a bicycle in most countries and is well defined in the

The Handbook of Lithium-Ion Battery Pack Design. http://dx.doi.org/10.1016/B978-0-12-801456-1.00015-4

European Union. Under the EU directive (EN15194 standard) for motor vehicles, a bicycle is considered a pedelec if:

1. The motorized assistance that only engages when the rider is pedaling, cuts out once 25 km/h is reached, and
2. When the motor produces maximum continuous-rated power of not more than 250 W (Association Française de Normalisation, 2009).

Today, batteries for pedelecs are most often lead acid, but may also be NiCd, NiMh, or lithium-ion (most frequently Lithium-Iron Phosphate (LFP)). They range from 12 V, 24 V, 36 V, and in some cases 48 V and carry capacities between 250 Wh and 850 Wh of energy. Today, the largest number of bikes are using lead acid batteries with 12 V. E-bike batteries are typically mounted on the frame of the bike, under the seat, or under a rear storage rack. One of the key design criteria for batteries for e-bikes is that they be portable. Most batteries are designed to be removed so that they cannot be stolen and so that they can easily be charged in homes or offices.

The second category of e-bikes are the electric scooters or mopeds. Electric mopeds can be either low speed or high speed, and they can either have pedals or not. In many cases, electric mopeds use standard lead acid batteries, however, just like the e-bicycle, electric mopeds are converting to NiMh and lithium-ion. These offer a reduction in weight and higher energy densities which means greater electric driving range. The batteries in these applications are typically permanently mounted to the vehicle as compared to the pedelecs, which are removable. Electric scooters are seeing the biggest demand in the Asian countries, with countries such as China, Taiwan, and India adopting them in great numbers and offering consumer incentives to help increase the usage (Figures 1 and 2).

Figure 1
AllCell Summit® e-bike battery pack.

There is also a growing market for electric motorcycles in many markets of the world. In the United States and Europe, there are several manufacturers that are working on premium and performance electric motorcycles. Companies such as Zero Motorcycle, Brammo, Mission Motorcycles, and most recently even Harley-Davidson have announced the introduction of high-performance electric motorcycles with lithium-ion batteries ranging in size from about 8 kWh up to about 15 kWh in size (Figure 3).

Figure 2
Electrical moped during charging.

Figure 3
Zero motorcycle.

Batteries for these motorcycles are primarily lithium-ion due to the high energy and power demands and are most often custom designed for the application due to the packaging restrictions of the motorcycle. These batteries typically need to meet similar performance and safety requirements as automobiles based on their motorcycle classification and top speeds, which are not typically limited as they are in mopeds and e-bikes.

The next largest market for electric-based personal transportation is one that is been around and is already selling in high volumes—the electric golf cart. The electric golf cart typically uses traditional 12 V lead acid batteries as this is a very cost sensitive market. Companies such as EZ Go and Club Car are major manufacturers of the golf cart-type EV. However, this market is also evolving to include some more premium applications such as electric versions of the traditional "four-wheeler" off-road vehicle. One example of this is the Bad Boy Buggy company which offers a complete line of both electric and hybrid electric "4 × 4" vehicles. These vehicles are mainly using lead acid batteries today, but are looking at solutions based on lithium-ion as well. Most batteries for these types of systems range from 48 V up to about 72 V.

Yet another category of electric personal transportation is the light electric vehicle (LEV). The e-bikes, scooters, electric golf carts, and 4 × 4s fall within this category, but there are also many other types of vehicles. The LEV is typically constrained by US state laws to a maximum road speed of between 25 mph and 35 mph and in some cases is limited only to access on certain roads. In most all cases, the LEV is not permitted for use on public highway systems. Some of the LEV's may be either three-wheel or four-wheel variety. The three-wheel vehicles are typically classified as motorcycles in most areas, so the same regulations and laws that apply to motorcycles apply to that category of LEV. One example of this is the Piaggio, a three-wheeled electric motorcycle that is produced by one of the largest producers of two wheel and commercial personal transportation.

Another example of an LEV is the Aptera 2e, which was a somewhat short-lived program but which offered a very unique product. The Aptera was based on a motorcycle structure, but included a body based on an airplane like aerodynamic design in a three-wheeled configuration. The company intended to include both series hybrid and fully electric versions of the vehicle. The fully EV was intended to include a 20 kWh lithium-ion battery, which at the time it was announced, intended to use an A123 battery pack until the company ultimately filed for Chapter 11 bankruptcy in 2011 (Voelcker, 2011) (Figure 4).

Other personal transportation applications that are currently electric powered include things like electric wheelchairs that typically use lead acid batteries and the Segway which uses a lithium-ion battery to provide short-distance transportation and is seeing significant use in areas such as airports, shopping malls, college campuses, and similar locations.

Figure 4
Aptera, with its revolutionary Typ-1, is radically restyling passenger vehicles to save weight and energy. Although classified as a motorcycle, Aptera has targeted exceeding passenger car safety standards in its design.

Looking further into the future, companies such as General Motors, Toyota, Honda, and others have begun working on personal transportation projects that are focused on meeting the challenges of the megacities of the future. Toyota introduced a concept car at the 2013 Geneva auto show they called a personal mobility vehicle called the "i-Road." The i-Road is a fully electric two-person vehicle with a range of about 30 miles (50km) (DesignBoom, 2013). While Toyota has not announced any production timing around this vehicle, it offers a glimpse at what the future of transportation could look like.

Similarly, at the Geneva Motor Show in 2010 Honda introduced the 3R-C Concept vehicle. This concept vehicle is a single person, three-wheeled vehicle intended for use in urban environments. In this case, Honda built on their significant experience in motorcycle development offering an open-air driving experience but with a glass canopy that will cover the driver's compartment when the vehicle is parked or not in use. Again, no production date has yet been shown but instead a vision of the future of personal transportation has been offered (Honda, 2010) (Figure 5).

There are many other examples of this type of product but I will offer just one more sample, the General Motors and Shanghai Automotive Industry Group (SAIC) introduced the jointly developed EN-V concept car at the Shanghai Auto Show in 2010. The EN-V is a two-person "pod"-type vehicle based on a "Segway" like propulsion system and powered only by lithium-ion batteries with a range of about 25 mile (40km) (General Motors, 2010) (Figure 6).

Personal transportation is expected to be a major area of growth for the electric transportation industry, with more than 60% of the world's population living in not only urban areas by 2030,

Figure 5
Honda 3R-C Concept vehicle.

Figure 6
GM EN-V concept.

but in many cases in the megacities of the future (Rhodes, 2014; United Nations, 2014). In these megacities, with populations in the tens of millions, challenges of parking traditional vehicles, high levels of emissions, and major traffic will increase the demand for new products to meet the challenges of these supercities.

Automotive Applications

Electric batteries for automotive applications will fall into four basic categories: (1) micro hybrids (uHEVs), (2) hybrid electric vehicles (HEVs), (3) plug-in hybrid electric vehicles (PHEVs), and (4) battery electric vehicles which may also come in the form of an extended-range electric vehicle (EREV) or even fuel cell EV. Each of these categories was briefly covered in Chapter 3, however, in this chapter will discuss some specific examples.

Microhybrid Electric Vehicles

The uHEV is a category that has grown significantly over recent years due to continually increasingly demanding CO_2 reduction and fuel economy standards. In Europe, this form of electrification has already made significant penetration into the market with more than 50% of the new vehicle sales having some form of uHEV technology.

There are basically two types of uHEV, a 12-V system and a 48-V system. Most of the 12-V systems tend to use lead acid batteries but are beginning to shift toward lithium-ion battery solution for some applications. Most of these applications are designed with about 250 Wh of energy. These applications do not provide any other support functionality to the vehicle except for the stop/start functionality (Figure 7). In order to get more electric functionality, system designers are moving toward 48-V-type applications. These batteries may have between 500 Wh and 1 kWh of energy. The biggest difference between the two is that the 48-V system will typically include the capability to capture regenerative braking as well as providing some minimal acceleration support, powering the A/C compressor and the auxiliary systems power. Many forecasts expect to see global market penetration of around 20% by the year 2020, other forecasts show penetration in Europe and the United States even achieving upward of 70–80% uHEV penetration by 2020.

There is not yet a set of standards for batteries of this type, rather they are being designed either specific to the vehicle or are being designed to fit into the footprint of a standard lead acid battery. One example is from Johnson Controls who have developed an "off-the-shelf" standard 48-V-battery system designed for use in stop/start applications. Johnson Controls has not yet published a data sheet for this yet, but based on their cell technology we can estimate that it likely has 13 cells in series and a single parallel grouping. And assuming it is using either their PL6P or PL27P means that it will have either 288 Wh (48.1 V × 6 Ah) or 1.3 kWh (48.1 V × 27 Ah) of energy on board (Figure 7).

Another example of a manufacturer who has adopted stop/start technology is Mercedes who offers a 12-V-stop/start technology using a lead acid battery on most of its vehicle line in Europe today, including the B-Class, C-Class, CLA, CLS, E-Class, G-Class, GL-Class, GLA, GLK, M-Class, S-Class, SLA, SLK, and the SLS AMG.

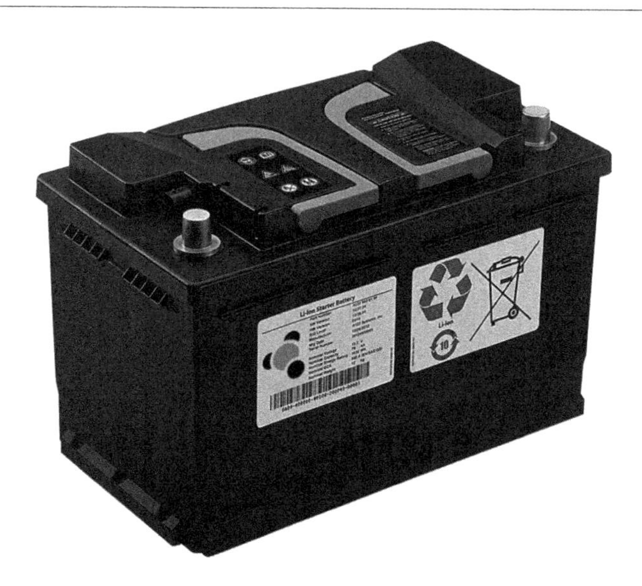

Figure 7
A123 Li-ion starter battery.

Hybrid Electric Vehicles

The next application of lithium-ion battery technology is the HEV battery, which can actually be broken down into two categories: mild hybrid and strong hybrid. The mild hybrid typically has lower system voltages of around 110–250V, while the strong hybrid has a system voltage in the range of 330–350V. Both mild and strong hybrids in their first generations used NiMh batteries but many are beginning to move toward lithium-ion solutions.

General Motors introduced a mild hybrid in their 2010 Chevrolet Malibu based on a Cobasys NiMh battery. In this case, Cobasys developed a 36-V air-cooled battery that was mounted in the trunk of the vehicle. In their second generation, GM transitioned away from NiMh and instead installed a lithium-ion battery. The new lithium-ion battery increased the voltage of the system up to 110V and offered some additional functionality that the first generation did not, including acceleration support and enhanced regenerative braking (Figure 8).

The best example of a strong hybrid is the Toyota Prius. It is the most popular and the highest volume hybrid on the market today. It has used NiMh batteries since its launch and continues to only use NiMh with the exception of their PHEV, which uses lithium-ion. The Prius HEV battery was developed using NiMh batteries, in this case 168 cells assembled into 28 modules to provide about 201V at the pack level. With a 6.5Ah NiMh cell, this offers about 1.3kWh of total energy. Toyota's pack integrates air cooling, using an internal fan to pull fresh air into the pack and exhausting it out the other end. The Prius HEV battery weighs about 93 pounds

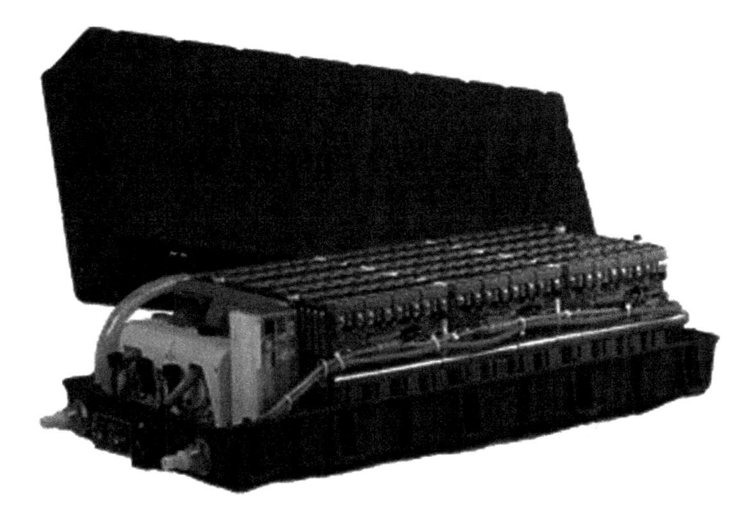

Figure 8
Cobasys NiMh battery.

Table 1: Toyota Prius battery specification (Toyota Prius Battery.com, 2013)

	1997 Prius (Gen I) Japan Only	2000 Prius (Gen II)	2004 Prius (Gen III)	2010 Prius (Gen IV)
Form factor	Cylindrical	Prismatic	Prismatic	Prismatic
Cells (modules)	240 (40)	228 (38)	168 (28)	168 (28)
Nominal voltage	288.0V	273.6V	201.6V	201.6V
Nominal capacity	6.0Ah	6.5Ah	6.5Ah	6.5Ah
Nominal energy	1.728kWh	1.778kWh	1.31kWh	1.31kWh
Specific power	800W/kg	1000W/kg	1300W/kg	1310W/kg
Specific energy	40Wh/kg	46Wh/kg	46Wh/kg	44Wh/kg
Module weight	1090g	1050g	1045g	1040g
Module dimensions	35(oc) × 384(L)	19.6 × 106 × 275	19.6 × 106 × 285	19.6 × 106 × 285

and mounts underneath the rear seat of the vehicle. In the table below (Table 1), a comparison of the Prius NiMh battery is shown.

From this you can clearly see how the design has advanced with the pack voltage being reduced from its original 288V down to just over 200V. This was done with a reduction in cells, again reducing from 240 cells in the introductory model down to just 168 cells in the 2010 model. Additionally, Toyota moved from a cylindrical NiMh cell to a prismatic can cell in the second generation of the pack and has remained with that design since. With the great success of this technology, Toyota has branded it the "Hybrid Synergy Drive" technology and has now expanded the product lineup to several other vehicles and brands (Table 1).

PHEVs and EREVs

The next type of automotive battery is the PHEV which I consider to be one of the most important, even if it is transitional, technologies available today. In the PHEV, a battery ranging between about 7 kWh and 16 kWh is most often used in addition to the internal combustion engine. The downside is that the vehicle carries the cost of two complete powertrains, one electric and one gasoline (or diesel). The basic concept with this technology is that the electric battery and electric powertrain will offer between 10 and 40 miles of all-electric driving range and once that initial electric range is exhausted the internal combustion engine will engage and provide continuous power to the motors. At this point, the vehicle begins operating more like a traditional hybrid rather than either Internal Combustion Engine (ICE) or electric thereby continuing to offer improved fuel economy over a traditional ICE. This combination offers the vehicle driver about the same total driving range as they would be able to achieve in a traditional ICE-powered vehicle (Figure 9).

Different OEMs have taken different stands on how much energy and therefore electric range to include in their applications. Several different studies have concluded that about 80% of all American drivers commute less than 40 miles per day, so this is where we see most of the OEMs focusing their PHEV efforts—on providing electric ranges up to but generally not above 40 miles. Toyota, for instance, has decided on a relatively short-range electric driving offering only 10 miles of electric drive range. Whereas General Motors has opted for a nearly 40-mile-range PHEV. In between those two, Ford is opting for about 20 miles of electric drive range. So depending on your daily commute, there is a solution out there to meet your needs.

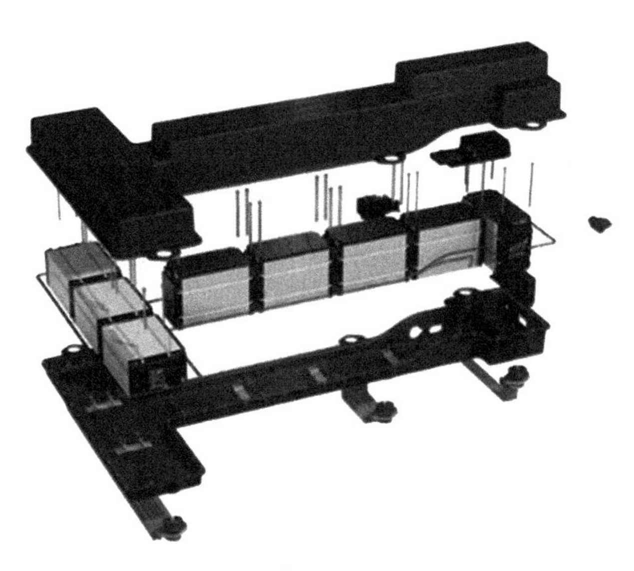

Figure 9
A123 PHEV lithium-ion battery.

Now here you may notice that I am combining two categories into the same discussion. In this instance, I am consolidating both traditional PHEVs with their closely related cousin, the EREV. In these cases, the main difference is whether the system is a parallel design or a series design. Both of these configurations were discussed in detail in Chapter 3, however, it is worth a brief review at this point.

In the parallel hybrid configuration, the electric motors are traditionally sandwiched into the drivetrain putting the electric motors in the direct driveline, whereas in the series configuration the ICE is not actually connected directly to the transmission and will operate like a generator providing power to the electric motors. The EREV essentially is always in EV mode, but with an on-board generator to provide continuous power to the wheels.

So now let us take a look at a couple of examples of these technologies in their applications. First, we will start with the Toyota Prius Plug-In Hybrid. The Prius PHEV uses a 4.4-kWh lithium-ion battery supplied by the Prime Earth Electric Vehicle (PEVE) company, which is a joint venture between Toyota and Panasonic (Green Car Congress, 2009). This battery offers the driver between 10 and 15 miles of electric drive range depending on factors such as the temperature, driving style, and other factors. With a relatively small battery, it can also charge very quickly, using a 240-V charger the battery can fully charge in only 90 min and with a standard 120-V charger in only 3 h. The battery weighs only about 180 pounds and mounts under the rear cargo floor area of the hatchback. According to the Environmental Protection Agency (EPA) ratings, the Prius PHEV achieves about 95 MPGe and uses about 29 kWh per 100 miles.

In the United States, Ford Motor Company has also taken up the PHEV race offering several different vehicles with the same PHEV battery technology. First is the Ford C-Max Energi PHEV followed by the Ford Fusion Energi PHEV and the Escape PHEV that was originally being planned for a 2012 introduction but has not yet hit the showroom. One aspect of the Ford battery strategy that has been seen in multiple conference proceedings and in personal conversations with the team is the drive to use identical batteries for all of these applications. Even talking about common designs is not sufficient to meet the directive. In order to achieve significant cost reductions, Ford's focus (pun intended) is to use the same battery across as many applications as possible. This will generate higher volumes and should therefore help to drive down the cost of the batteries faster.

Both the Ford C-Max Energi and the Ford Fusion Energi PHEVs use a 7.6-kWh battery pack that is supplied by Panasonic. This offers a pure electric drive range of about 20 miles and an EPA estimated 88 MPGe using about 37 kWh per 100 miles. In the C-Max, the battery mounts under the rear cargo floor. The battery is air cooled, pulling fresh air in through an integrated fan and exhausting out the top of the battery pack in two locations. The biggest challenge with Ford's requirement to use identical batteries is that the battery packaging is not specific for each vehicle therefore the packaging does not necessarily fit nicely into all of its applications (Figure 10).

Figure 10
Ford C-Max lithium-ion battery pack. *Courtesy Ford Motor Company.*

General Motor's Chevrolet Volt is an example of the sister technology to the PHEV, the EREV. Combining a 16.5-kWh lithium-ion battery pack using LG Chem pouch-type cells with a small internal combustion engine, the Volt offers the driver a range of about 38 electric miles and achieves an EPA rating of about 98 MPGe using about 35 kWh per 100 miles. The Volt's battery pack is mounted underneath the vehicle filling the former transmission tunnel and fuel tank areas in a "T"-shaped pack. GM went with a liquid-cooled solution to help ensure both rapid cooling and in colder temperatures faster heating of the batteries. The Volt pack, branded "Voltec" by GM uses a total of 288 lithium-ion pouch-type cells assembled into four modules. Each cell is separated by a plastic frame on one side and an aluminum cooling fin on the other side. The cells are electrically connected on the top of the module using a plastic interconnect board with the series/parallel connectors integrated into that board. The Voltec pack also includes a service disconnect in the middle of the pack that is accessible from the passenger compartment. To help manage the pack's temperatures, a heat shielding insulation covers the top plastic cover (Figure 11).

Battery Electric Vehicles

Perhaps the most well-known fully EV on the market today is made by Tesla. Their first introduction was the Roadster and their second vehicle is the Model-S sedan. The fully EV differs from the hybrids in that 100% of the power and propulsion are provided by an electric battery, there is no ICE to back up the vehicle in these applications.

In developing their battery solution, Tesla went a very different route than the "big" OEMs. Rather than creating a new battery format that fits in the into the vehicle architecture, they

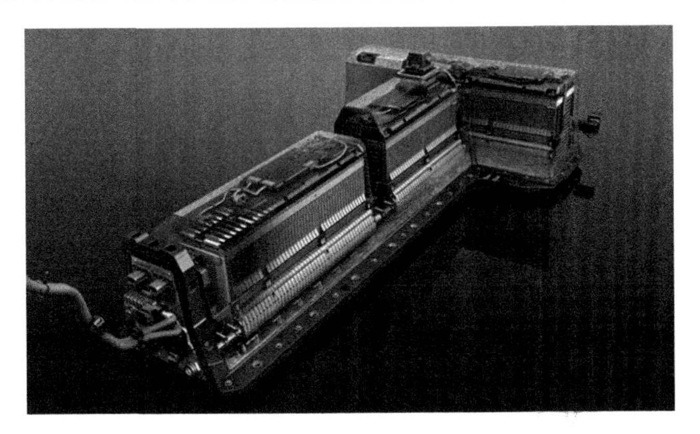

Figure 11
2012 Chevy Volt lithium-ion battery pack.

opted to use the only standard lithium-ion form factor that is already being produced in massive volumes annually—the cylindrical 18650 (18 mm diameter × 65 mm height) as the basis for their packs. The belief was that by using an already commoditized battery technology it will help to make the final vehicle price more affordable. And in an application that offers an 80-kWh battery pack, the battery is by far the biggest cost driver in the vehicle so managing the battery cost is crucial.

In the Roadster, Tesla developed a pack using 6,831 18650-type lithium-ion cells. The cells were built into 69 cell "bricks," nine of those bricks were then connected in series to form a "sheet" and 11 sheets connected in series form the completed battery pack. Tesla claims to have developed the pack to prevent cascading and propagation of failures during thermal runaway events, however, they have not disclosed exactly how they are doing that and several recent events may question how well those solutions are working. The Tesla Roadster pack uses a liquid-cooled design, constantly circulating liquid throughout the pack even when charging and mounts behind the seats in the storage area of the vehicle (Figure 12).

Tesla's second offering, the Model-S, continues to evolve the technology by growing the battery pack to 85 kWh, using 7104 lithium-ion 18650-type cells. The Model-S battery differs from its predecessor in that it is designed to be extremely flat, mounting underneath the vehicle. It is only about 6-inches deep and it spans from the rear axle to the front and from one side of the car to the other. While Tesla did not name the source of the batteries for the Roadster, Panasonic was identified as the source of the lithium-ion cells for the Model-S, which uses an NCA (Nickel Cobalt Aluminum)-based chemistry. Tesla offers both a 60-kWh and an 85-kWh battery for the Model-S (Figure 13).

The Renault-Nissan group has also been introducing multiple new fully EVs, perhaps the best known and the highest volume selling EV on the market is the Nissan Leaf. The Leaf uses a 24-kWh battery designed by Automotive Energy Supply Company (AESC). The lithium-ion

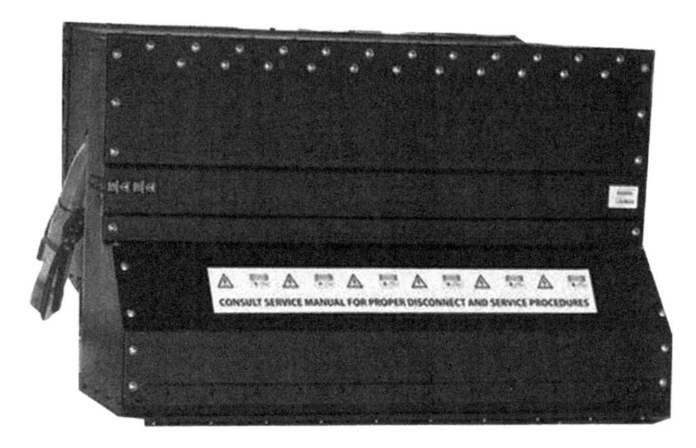

Figure 12
Tesla Roadster lithium-ion battery pack.

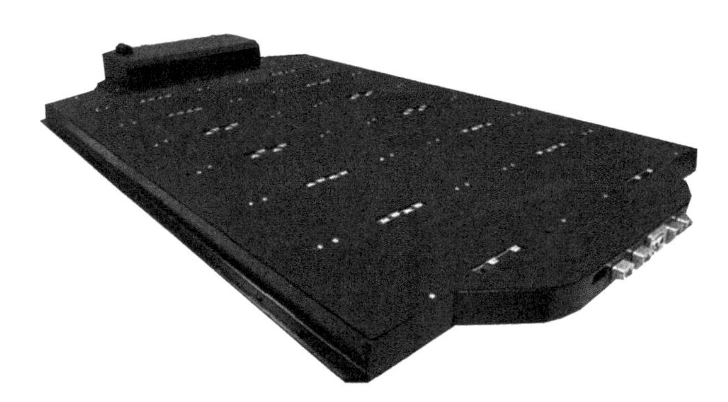

Figure 13
Tesla Model S lithium-ion battery pack.

battery uses a total of 192 pouch-type cells that are installed into four-cell modules. Each module includes two cells in series and two in parallel. The modules are then stacked and mounted in the battery enclosure. Each module is essentially sealed, with no active thermal management system installed in the pack. In this application, the thermal management is being done via passive means, with the heat of the cells being transferred to the metal enclosure of the modules and then to the external pack enclosure. The Leaf also uses a centralized battery management system, with a single control unit and a wiring harness that extends to each module. Mounted external and underneath the vehicle, the pack is also sealed in order to prevent dust or liquid intrusion (Figure 14).

Nissan's sister company, Renault has also launched several fully EVs including the Renault Twizy, a small two-person city car with a 6.1-kWh lithium-ion battery; the Renault Zoe a mid-size four-door sedan with a 22-kWh lithium-ion battery; and the Renault Kangoo Van

Figure 14
AESC battery module for Nissan Leaf.

with a 22-kWh lithium-ion battery. One of the interesting things that Renault has done with their EVs is to offer the battery as a separate monthly service agreement in order to keep the vehicle costs low. Initially, all of the Nissan and Renault EVs used the AESC batteries, however, in a recent announcement the Renault-Nissan group will begin looking at external battery companies in an effort to continue to drive commoditization and cost reductions in the batteries (Ciferri, 2014) (Figure 15).

Ford has also introduced a fully EV in the Focus EV. The Focus EV was initially integrated from ICE to electric by Magna E-Car using a 23-kWh liquid-cooled battery designed by LG Chem. The battery is somewhat unusual in this application in that it is

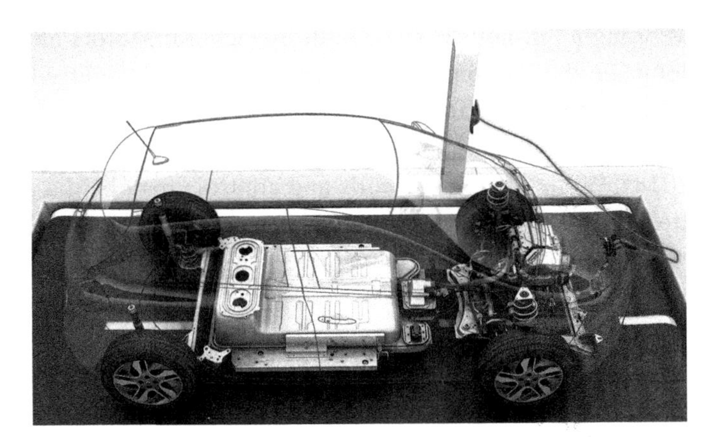

Figure 15
2013 Renault Zoe electric vehicle. *©2012-2013 Nissan North America, Inc. Nissan, Nissan model names and the Nissan logo are registered trademarks of Nissan.*

Figure 16
Ford Focus electric vehicle chassis and lithium-ion battery. *Courtesy Ford Motor Company.*

split into two separate boxes. One mounts under the rear seat, the second mounts in the trunk. By using a liquid-cooled thermal management system, Ford is able to manage the temperatures of the two battery packs very closely. Additionally, by using a liquid-cooled solution they are able to both heat the battery in the winter as well as cool it in the summer time (Figure 16).

General Motors has also introduced an all-electric small car called the Chevrolet Spark. The Spark initially used a 21.3-kWh-nanophosphate lithium-ion battery that was designed by A123. However, with the 2014 model year vehicle the battery pack was entirely redesigned by LG Chem's US subsidiary Compact Power. The new battery assembly was also brought "in-house" to be built by General Motors at their Brownstown assembly plant alongside the Volt battery. The new battery has slightly lower capacity, about 19 kWh but claims the same driving range as the original. This change has allowed GM to commonize on both cell and module designs in order to help drive down battery costs (Voelcker, 2014). While the battery cells and controls may be the same, the pack assembly was entirely redesigned in order to package the batteries in the Spark. The lithium-ion batteries mount underneath the vehicle beneath the rear seats and under the rear storage area (Figure 17).

Fuel Cell EVs

Another potential game changer in the energy generation and transportation industry is the fuel cell. A fuel cell uses a polymer electrolyte membrane also called proton exchange membrane using hydrogen as a fuel source and oxygen from the air to produce electricity.

Figure 17
Chevrolet Spark electric vehicle.

They are in effect electricity generators, however, they often need a relatively small amount of on-board energy storage in order to ensure a constant supply of electricity which makes the fuel cell vehicle an EV. By integrating an energy storage system with the fuel cell, it may be possible to reduce the size of the hydrogen storage tank by adding a larger battery. The battery also enables the vehicle to recover energy through regenerative braking.

In the automotive space, most of the major automakers have had ongoing development in the fuel cell vehicle application. Honda has introduced the Honda FCX Clarity fuel cell vehicle that integrates a 288-V lithium-ion battery into the vehicle. Similarly, Hyundai has introduced a Tuscon fuel cell vehicle with a 188-V lithium-ion battery. General Motors introduced a demonstration fleet of Chevrolet Equinox's equipped with their latest fuel cell technology and have announced plans to introduce a production version of the technology in 2015/2016 time frame (Cobb, 2014). Toyota, BMW, Mercedes-Benz, Mazda, Fiat, Audi, Nissan, and Volkswagen have all been working on fuel cell vehicles as well.

In addition to the vehicle applications, fuel cells are being demonstrated in large energy storage applications worldwide and have even made significant progress in places like Japan where small household-sized fuel cells are regularly being used (Japan Echo Inc, 2003; Runte, 2014). However, the household units do not typically require a battery as they are used solely for power backup.

While many manufacturers are working on fuel cell solutions, it is likely to continue to be some time before they become viable for the mass market. While the technology is evolving rapidly, there is still a major issue in terms of hydrogen supply infrastructure. Until there is a reliable source of hydrogen that is publicly available, fuel cell vehicles will struggle to gain significant market share.

Bus and Public Transportation

Another interesting area that is experiencing massive growth in electrification is the realm of public transportation. There are many drivers for this growth, including the demand to improve air quality in urban areas, reducing fuel costs, and improving rider experience. And like much of what we have discussed in this book, electric buses and light rail systems are not an entirely new idea. In cities such as New York and San Francisco during the early twentieth centuries, most of the public transportation was provided either by electric streetcars and for a short period from the late 1890s to the early 1920s also from electric buses (Kirsch, 2000).

The one factor that is beginning to become clear to fleet operators is the reduction in fuel costs that can be achieved through electrification. If you consider that the typical diesel-powered bus gets between 3 and 6 miles per gallon depending on vehicle size, route (cycle), and the fuel type, that makes fuel costs just about the biggest expense that the municipality will incur on an annual basis (after the initial capital expense). Reports (Clark, Zhen, Wayne, & Lyons, 2007) have indicated that the average municipal bus will operate for just under 40,000 miles per year. If we do some simple calculations based on this, we find that the average bus will burn about 10,000 gallons of diesel fuel per year (assuming 4 mpg) which then equates to about $38,430 USD in annual fuel costs for a single bus. If we assume that the municipality has at least 10 busses in their fleet, then the annual fuel costs will be about $384,300 USD. By the addition of either a hybrid, plug-in hybrid, or even an electric battery system into the vehicle, we could expect to achieve about 20% fuel savings for a hybrid and up to about 50% for a plug-in hybrid system. That means that the operator of that fleet could see savings for a plug-in type hybrid system of up nearly $200,000 annually or nearly $2 million over the 10-year life of the fleet.

Additionally, as we continue to see increases in major urban population centers, which are also experiencing major increases in pollution due to internal combustion engine-type vehicles, there are increasing demands for greater amounts of electrified public transportation. Everywhere in the world this demand is centered on the major urban cities such as San Francisco, New York, Beijing, Shanghai, Tokyo, London, Paris, and many other major cities as they evolve into megacities.

Modern examples of the electric bus include companies such as Proterra who is developing an all-electric bus that is designed from the ground up to be an electric alternative to traditional powered buses. Companies that are developing electric and hybrid buses include major systems integrators such as BAE Systems and Allison Transmissions, system integrators and major manufacturers such as New Flyer, Gillig, BYD, Wrightbus, Volvo, Novabus, Yutong, Lian Fu, Zhong Tong, and many others are all developing electrified bus applications.

The battery for an electric bus will vary depending on a couple of considerations. First, is it a fully electric bus or is it a hybrid or plug-in hybrid electric-powered bus? For an all-electric bus, we should then ask whether or not it is expected to fast charge during its operating cycle? If the bus is expected to charge frequently during its operational shift, it is important to understand the capability of the lithium-ion battery cells to accept frequent rapid charging, otherwise the life of the battery could be prematurely limited. However, by including frequent in-service charging opportunities it may be possible to reduce the overall size, and therefore, cost of the battery. The overall size of the battery will vary depending on the route, usage, and specific requirements but an all-electric bus could have a battery anywhere from 75 kWh up to 300 kWh and depending on the vehicle electrical architecture design will be at least 350 V but more likely will operate in the 650 V range. One other aspect should be mentioned here, the size of the battery is related to the distance the bus needs to travel. For instance, if the bus integrates a battery that can accept frequent fast charging, the battery can be of smaller size, both physically and energy-wise. This means a lower cost and lighter battery can be used in place of a larger, heavier battery. However, this is only effective if the battery chemistry can accept frequent fast charges without degrading the performance over its life (Figure 18).

Hybrid buses include an electric battery in addition to either a liquid propane gas (LPG), diesel, fuel cell, or other propulsion engine. In this case, just like in the automobile hybrid, the battery will be much smaller than a fully electric bus battery but will likely still be much larger than an automotive-type HEV battery.

In addition to buses, much of the rail transportation includes some form of electrification. For many years, large diesel-powered locomotives have integrated large capacity lead acid batteries to create diesel-hybrid locomotives. Today, some reports indicated that as much as 50% of the global light rail is powered by electricity. Some of these are integrating lithium-ion batteries in order to reduce emissions and improve performance. There are also some very

Figure 18
New Flyer Xcelsior electric bus.

unique applications where battery electric rails make the most sense such as in mining applications, underground tunnels, and in urban public transportation.

One final thought on the electrification of commercial transit vehicles. While safety is important in every lithium-ion battery application, in those such as a public bus which will carry many people at once, safety is absolutely vital. In this instance, a failure must be able to be contained outside of the passenger compartment. Even better in this case would be a battery pack design that can isolate a failure to a single cell.

HD Truck Applications

Over the road and heavy-duty trucking is also experiencing increased demand for electrification. With new regulations being enacted annually that limit engine idling, there is greater need for increased amounts of electrification to provide housing loads, increase fuel economy, reduce fuel demand for trailer cooling, and similar applications.

In fact, many US states have already enacted "anti-idling" regulations that do not permit semitrucks and other large trucks (Gross Vehicle Weight Ratio (GVWR) 10,000 lbs or heavier) to sit idling for longer than 5 min (California Air Resource Board, 2013). This means that these trucks can no longer use engine power to run their "housing" loads to power their environmental control (heating and cooling) and other auxiliary power devices (computers, TVs, etc.) while the truck is at rest. Many of these vehicles are now choosing to install battery power to run these devices while the truck sits overnight or for long periods.

Some of these applications are relatively small, less than 5 kWh and must be very light and compact. The "best" battery in this instance would be one that can work both as a high-power starting load for the engine as well as an energy source to operate the housing loads. This may be accomplished through a battery system that includes both lithium-ion and another power source such as lead acid or an ultracapacitor, or in some cases even with advanced lead acid solutions. But by using a combined "hybrid" battery the lithium-ion can supply the energy needs of the system, while the lead acid or ultracapacitor provides the power needs for starting the engine.

Other HD truck applications are looking to use battery power to maintain the temperatures of their refrigerated trailers as well. Similarly, these trucks have historically used small diesel engines to run the refrigeration units of their trailers. Today they are working to install battery power of about 40 kWh to 50 kWh in order to maintain the cold temperatures needed to haul frozen products.

Industrial Applications

Batteries and energy storage also provide power for a wide range of industrial applications. If you go into many large warehouses, you will see that the forklifts and other equipment used to transport material throughout the facility are currently being powered by lead acid batteries. But both lead acid batteries and LPG are beginning to be displaced by lithium-ion batteries. The value proposition for replacing lead acid is fairly easy—lithium-ion can provide longer operating times. Many large warehouse vehicles need to replace their batteries at least once per shift. That is effectively nonproductive down time when the equipment is not providing any service to the operation. By transitioning to a lithium-ion battery, many of these applications can operate for a full 8–10 h shift without having to replace the battery. One thing to keep in mind when it comes to electric forklifts is that often they need the weight of a lead acid battery to act as a counterbalance to the weight of the material they are moving. In these situations, lithium-ion again may not be the best current technology but can still find some very good applications (Figure 19).

As mentioned previously, the largest single fleet of EVs in the world is the golf cart. Today these applications are very cost sensitive and lithium-ion does not make a very cost-effective solution. However, the lithium-ion batteries of tomorrow or the second-life lithium-ion batteries may change this. While the golf carts themselves may not be a great fit for lithium-ion batteries today from a purely cost perspective, there are many associated vehicles with very similar designs that do make for a good fit for lithium-ion batteries. Examples of this include shuttle buses that carry people around campuses, airports, and industrial sites. Personal on-road and off-road transportation are also gaining much interest in lithium-ion batteries. In many communities, it is not uncommon to see people roving around their neighborhood in small, personal electric transports that are based on the same technology as the golf carts.

Figure 19
Crown electric forklift.

If you have done any air travel you may also have seen examples of electrically powered equipment used in a variety of applications on the airport tarmac, from the equipment that hauls your luggage to the plane to the people movers inside the airports that help people get around. These are both examples of equipment that either is or can be powered by lithium-ion batteries.

Robotics and Autonomous Applications

Another growing industrial use for battery power is in automation, robotics, and unmanned vehicular applications. Companies like iRobot have developed a full line of automated robotic applications ranging from floor vacuums, floor scrubbers, mopping, pool cleaning, and gutter cleaning. In addition, they have developed a whole line of robotic products for military and police use (Figure 20).

Other companies are working on automated robotic solutions for industrial warehouse use. In 2012, Amazon acquired the robotic company Kiva Solutions for the purpose of automating the Amazon warehouses (Greenfield, 2012). With that announcement, Amazon indicated their shifting warehousing strategy toward fully automated warehouses that will help to reduce costs as well as response time for their customers. Their current batteries allow them to operate for about 8 h and travel between 13 and 15 miles before needing to return to an automated recharging dock.

Due to the compact sizes and low cost targets for battery packs in these applications many of the battery solutions are based on 18650-type cells and operate in the range of 7–14 V. Today, many of these are using NiMh and NiCd cells, but are beginning to make the transition to lithium-ion.

Figure 20
iRobot Roomba robot vacuum cleaner.

Another burgeoning market for lithium-ion batteries are the autonomous vehicles. These can range from autonomous underwater vehicles (Figure 21) and unmanned underwater vehicles to unmanned aerial vehicles and remote-piloted vehicles. Many of these are being used by research, military, and police organizations as water-, land-, and air-based drones for reconnaissance and intelligence gathering operations. There are also early discussions about using these types of applications in support of public safety but currently face significant questions about personal privacy rights.

Marine and Maritime Applications

A very interesting set of applications that are beginning to emerge are in the marine shipping and maritime applications. Now that does not mean to infer that there have not been batteries in marine applications previously, clearly submarines, submersibles, and many other applications have used battery power. However, today we see large ships, ferries, tug boats, offshore-vessels, platform service vessels, and many other vessels integrating both fully electric and hybrid electric systems into both new and retrofit shipping applications.

One example of this is a Scandlines ferry based in Norway which has installed a 2.7-MWh-battery system designed by Corvus Energy and using XALT Energy lithium-ion cells in conjunction with power electronics systems designed and installed by Siemens. At the time of this writing, this is the largest hybrid electric ferry in operation in the world (Figure 22).

The benefits for this type of application can be huge. For the ferries and other ships that spend significant amount of time at port, there is the opportunity to "cold-iron" the ship, or to run the ship from dockside power while it is docked. This also offers a good opportunity to charge the battery which can offer significant reductions to fuel consumption of the vessel. For a

Figure 21
Autonomous underwater vehicle.

Figure 22
Corvus Energy marine battery array.

diesel-hybrid system, the vessel may be able to generate 15–25% fuel savings—which may not sound like much, but if you consider that these vessels may spend millions in fuel costs annually, this now begins to generate major cost advantages.

The other factor that is driving greater amounts of energy storage to be added are growing regulations around CO_2, SOx, and NOx emission and the drive to reduce them. One such regulation is the United Nations International Maritime Organization's "MARPOL" regulation which came into effect in January of 2013. These regulations are directed at reducing the emissions of SOx, NOx, and CO_2 from ships by 30% by 2030 (International Maritime Organization, 2014).

Grid and Stationary Applications

The market for stationary and grid-connected energy storage systems is quite large and encompasses a wide variety of different applications and storage technologies. Stationary applications can range from uninterruptible power sources that are used to provide backup power for data centers and other similar operations all the way to large multi-megawatt-size energy storage systems to provide power quality smoothing, integration of renewable energy sources such as wind and solar, and energy time-shifting functions. In between these two ends of the spectrum, there are a wide variety of other applications and functions for energy storage. These may come in the form of small household energy storage batteries that range from 5 to 15 kWh in size to community energy storage (CES)-type systems that can be used to shore up the power system at the neighborhood level.

The other emerging area of interest for stationary power systems is known as "islanding" which is becoming especially useful in remote areas or in areas where power reliability is an issue. Islanding occurs when a battery backup unit is installed, often in line with a generator, in order to provide uninterrupted service during outages of the regular power system. These systems may operate only for short periods of time, allowing the generators time to kick in and get ramped up, or they may last for hours at a time thereby reducing the overall fuel demand on the generator system.

One of the major drivers for these types of grid and stationary energy storage are the introduction of new regulations around the world. One such example is in California where the legislature has adopted new energy storage rules that require the State's main utility companies to put over 1325 MW of energy storage in place by 2020. This regulation was put in place in order to support another piece of State legislation that requires the same utility companies to generate at least 33% of their energy from renewable sources by 2020 (Sweet, 2013). This is the first such legislative mandate in the United States but will not be the last.

Utility-side grid storage solutions generally fall into one of several different categories, consumer-side energy storage solutions are either home-based or community-based energy storage solutions. One well-written source of information on the stationary grid-type energy storage applications is the "DOE/EPRI 2013 Electricity Storage Handbook in Collaboration with NRECA" (Akhil et al., 2013) which describes and defines the various types of applications for energy storage as well as describing the major types of storage that are being used in these various markets and defining how each one fits for each of the different markets (Figure 23).

Grid energy storage is interesting in that there are quite a few options when it comes to storing energy including compressed air energy storage (CAES), pumped hydro, flywheel, lead acid batteries, sodium–sulfur batteries, NiMh, NiCd, flow batteries, supercapacitors and ultracapacitors,

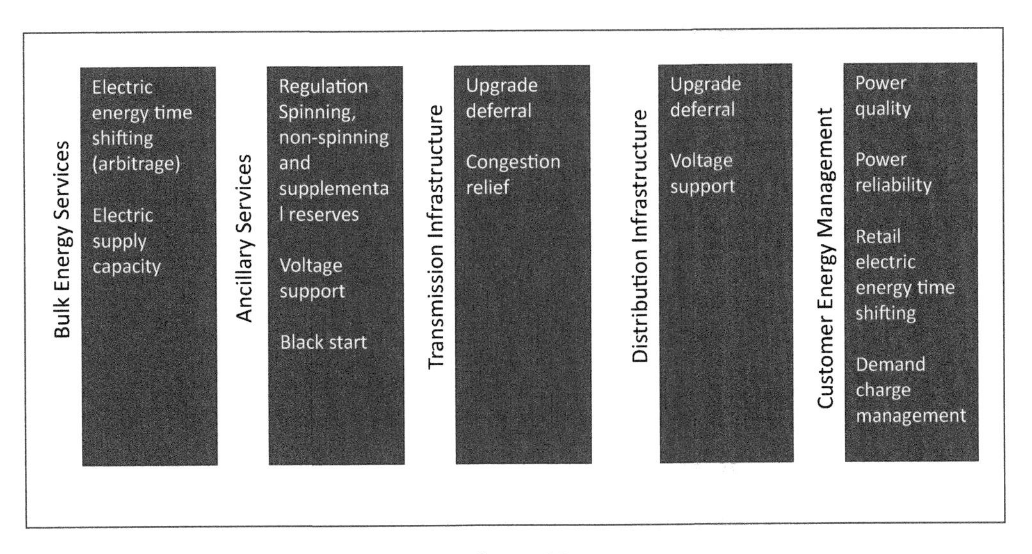

Figure 23
Grid-based energy storage markets.

fuel cells, and many other options. Each of these technologies offers different advantages and so they do not necessarily all overlap in their uses. For example, pumped hydro is an extremely cost-effective solution to storing energy but it is limited geographically and so is not a fit in every location. For our purposes, here we will focus on the lithium-ion battery-based solutions but will describe the basic storage applications that are using some form of energy storage.

And while we will discuss some of these energy storage uses individually, remember that the best case scenario is one that will use an energy storage system for more than one usage. This increases its value to the utility owner and operator and makes the value proposition much stronger for technologies that are capable of supporting multiple uses.

The graph below (Figure 24) shows a comparison of the major forms of energy storage in relation to their grid storage-type uses. The scale on the left side of the graph shows the time that the unit must be capable of discharging, ranging from seconds to hours and the scale on the bottom shows the amount of power that must be discharged during that period of time.

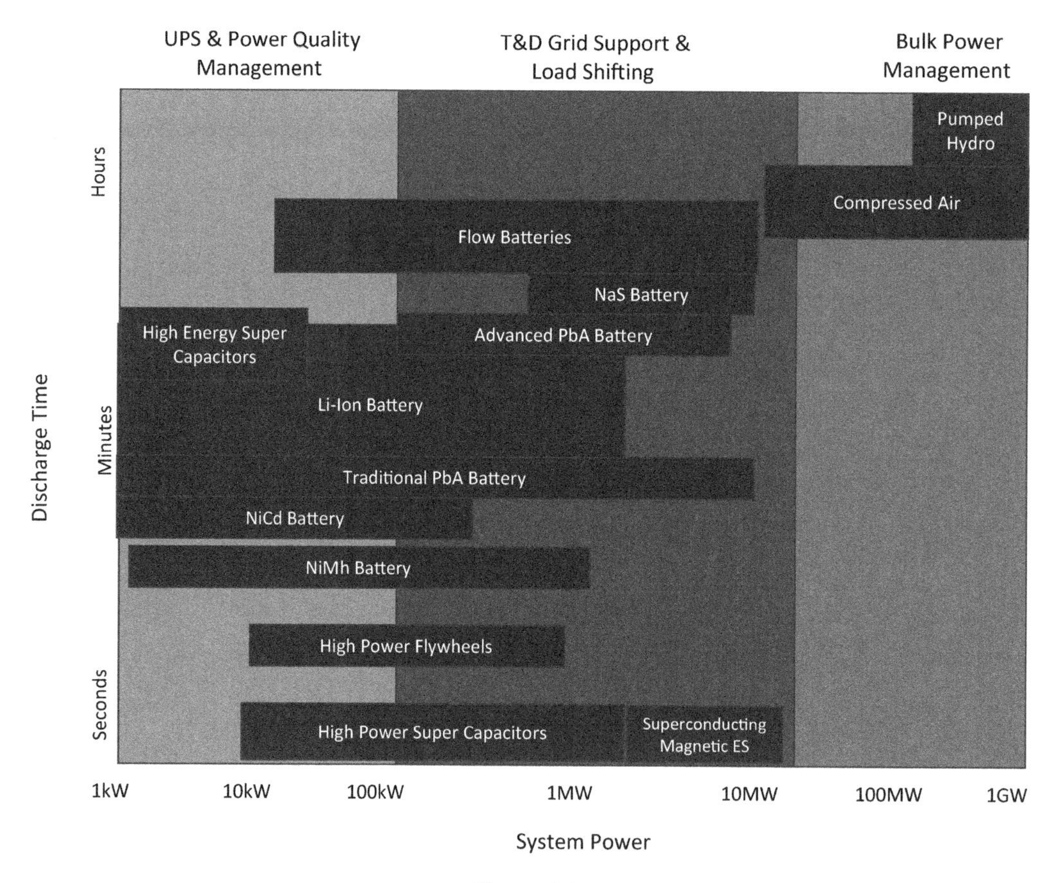

Figure 24
Types of energy storage for grid scale units.

There have been many forms of this same graph shown but the story is about the same, there are some energy storage technologies that are a better fit for certain uses—there is no "one size fits all" energy storage solution.

Bulk Energy Storage

Bulk energy storage includes time-shifting, arbitrage, and providing additional electric capacity supply. Time-shifting involves storing energy that was generated using an intermittent source such as solar or wind and using it when demand is high. This is especially useful as more renewable energy generating sources are brought online, and can be especially useful in island or remote regions where renewable generation allows the region to become more self-sufficient.

Energy storage for time-shifting applications has become very important because renewable energy is most often not generated at the time it is most needed. For example, solar energy generation typically peaks in the early afternoon, however, typical household usage generally peaks between six and seven o'clock in the evening. Similarly, wind generation tends to be greatest between midnight and the early morning hours—a time when it is least needed and also the time when energy usage is at its lowest and its cheapest. So by adding an energy storage solution to your renewable generation, you are able to store the energy until the time when it is most needed—in essence, you are shifting the energy generation to meet the customer demand.

Energy arbitrage involves purchasing energy at times when it is lower cost, such as in the middle of the night and using it during times of greater demand such as in the late afternoon when energy costs are at their highest. This is a major way in which the utility companies, both large and small, are able to generate increased revenues as well as reducing the need to add "peak power" to meet the daily peak energy demands. In almost every region, energy generation is the least expensive during the hours between midnight and six in the morning. This is the period when there is the least demand on the system and there is typically a lot of excess capacity. So utility companies can store energy generated during this period and use it when generation is at its highest cost.

The example above is a rendering that was created by A123 Systems to show what a large containerized grid storage system may look like (Figure 25). Companies such as A123, Amperex Technology Limited (ATL), EnerDel, Samsung SDI, XALT Energy, and many others have been or are developing solutions such as this with an all-in-one container that includes the batteries and controls, the climate control systems, safety features such as fire suppression, and access for system maintenance and controls.

Electricity supply capacity is simply the use of electric energy storage in order to avoid the major capital investments required of adding new coal or other high-capital energy generating capacity. More often utilities are turning to the integration of energy storage solutions such as batteries to avoid the high cost and emissions that would be generated by the traditionally powered facility. This is quickly becoming an important aspect of energy storage because of the major capital

Figure 25
A123 Grid Storage System™

expenses associated with putting in new power plants. Many utilities are experimenting with installing energy storage in their distributed networks and strategically located near the areas of greatest demand. When compared to the massive capital costs of putting a new coal-fired power plant in place many utilities are finding that it is much more cost-effective to install smaller energy storage systems strategically within their networks in order to either delay or offset the capital expenses that would otherwise have been generated by installing new power plants.

Ancillary Services

One of the most important ancillary services that are offered is frequency regulation. Frequency regulation involves managing the variation in generation and ensuring that the output meets the legally required frequency range. In the United States, the legally mandated frequency that energy must be distributed at is 60 Hz and in Europe it is 50 Hz. If the power that is being distributed falls too far outside of an acceptable range, often as low as 0.5 Hz from the mean, many utilities will automatically begin to disconnect in order to avoid damaging the equipment. Therefore, energy storage can be used in order to ensure that these "spikes" in frequency are regulated and that the power being transmitted is always at the specified frequency. Consider the battery a buffer in this instance being used to reduce the differences in demand and load and maintain a constant frequency.

Another ancillary service that may use energy storage is in support of spinning reserves and nonspinning reserves and is referred to as a supplemental reserve. In the United States, there are regulations that require utility companies have power generation capacity available that is in excess of current usage and can respond within 10 min. If this capacity is required to be online but not being used, it is referred to as "spinning reserves," this is in essence a power generation that is up and running and can respond quickly to any potential outages. In addition to this, there is a requirement for "nonspinning" reserves of power. This has the same requirement that it can be spun up and be online within 10 min. Finally, backup reserves are a backup power that is stored and that can be brought online after all of the other reserves are up and running. In essence, this system is designed in order to ensure that even during the peak usages, such as during the hottest days of the year, there is enough power to meet all demands. In this application, batteries are also beginning to be introduced in order to provide some of the supplemental reserves.

Another ancillary service is voltage support. This function operates in much the same way that frequency regulations services operate. Voltage support provides a regulation of the voltage of the system during times of peak demand and usage. Finally, black start is an ancillary service whereby energy storage is used as a backup to provide energy and power to the grid during catastrophic type failures of the grid system. In this case, the stored energy is used to power the grid until the main generators can be brought back online.

Transmission and Distribution Infrastructure Services

Transmission and distribution (T&D) upgrade deferral is the act of installing smaller energy storage systems throughout the utilities service area instead of installing new capacity generation equipment in order to meet peak demands. As was discussed earlier, most utilities are required by legal mandate to have installed capacity to meet the peak usages of the grid, even though those peaks may only occur as little as a couple of days per year. In this case, energy storage can be installed in strategically located areas in order to meet the demand on those days instead of installing expensive new main power generation equipment.

Another aspect of T&D deferrals and services is the use of energy storage in order to relieve the congestion in the system. This occurs when transmission sources are not adequate to support the peak demands on the system. In these instances, the utilities may actually incur additional transmission charges due to the congestion in the system as well as having to use higher cost energy sources during these periods. Again, strategically located energy storage systems can be used in order to relieve the stress on the T&D systems and to ensure continuous supply to the customers.

Customer Energy Management Services

Energy storage can also be used to improve the customer's power experiences. For instance, energy storage can be used to reduce the variations in the power and voltage, reduce the variations in the frequency, and to reduce the impact of interruptions in service to the end customer. Reliability can be ensured through the installation of energy storage as well. This essentially provides "islanding" support to the customer to ensure that there is no loss of service during catastrophic events.

Another customer service is being able to "time-shift" the energy usage. Many utilities are offering "time-of-use" billing, where the customer is charged a different rate depending on the time of day. For example, higher rates would be charged during peak times and lower rates during the late hours of the evening. In this case, the addition of an energy storage system on the customer side of the equation may allow the customer to store energy that is generated during the lowest cost periods and to use it during the most expensive periods. The net result of this is a reduction in the cost of energy to the customer.

There is still much debate as to whether the utilities or the customers should own the energy storage on the customer's side of the meter. The advantage of the utility owning this resource is that it would be able to manage many small energy storage systems as if they were one larger system, in essence, being able to send commands to discharge or charge these batteries at times of peak demand. On the other side of the coin, the consumers, if they have made the personal investment in these resources, may not be willing to give control to the utilities unless there are some financial incentives for them to do so.

Community Energy Storage

American Electric Power (AEP) has led the way in the development of a set of standards around CES energy storage systems. The CES is a form of distributed energy system that can be used to supplement a utilities capacity, add capacity in areas with greater demand, to create a microgrid, or to create an "island" of power. Typically CES units are developed in 25-kWh increments up to about 100 kWh and are installed in neighborhoods, often near a transformer (Figure 26).

Aerospace Applications

One other final area of interest in respect to lithium-ion and advanced battery systems is in the aerospace industry. Companies such as ABSL, Quallion, Saft, and Mitsubishi Electric have spent many years developing products for use in orbital satellites and other space-based applications. During the battery industry consolidation that occurred in the early 2010s, lead

Figure 26
Community energy storage unit.

acid battery manufacturer Enersys acquired both Quallion and ABSL in order to focus on the military and aerospace market for primary and secondary lithium-ion batteries.

The National Aeronautics Space Administration (NASA) has also been working in concert with several battery manufacturers to develop battery systems that can survive the challenging uses of space-based applications, from advanced robotics and space station power to lunar and Martian rovers.

Until the early 2000s, the main types of batteries used in satellite and space-based applications were either NiCd through the 1980s and then NiMh through the 1990s. It was not until the early 2000s when lithium-ion began being introduced into space-based applications. Today, more than 98% of all batteries used in government, private, and commercial space applications are lithium-ion based (Borthomieu, 2014). The biggest reason for the switch to lithium-ion batteries in the aerospace world is pretty clear—higher energy density. In these applications, managing the weight of the final product is vital, so by integrating lithium-ion batteries, engineers were able to increase the amount of energy on board while still reducing the weight. In satellite applications, there is a direct relationship between weight and cost. So the lighter the unit, the lower the cost to put it into orbit.

There are typically three different types of satellite applications and each one has a different battery requirement. First are the low earth orbit satellites. This requires about a 60-minute discharge cycle with a 30-minute charge period that is repeated multiple times per day as the satellite circles the earth. But they also have a lifetime requirement of up to 15 years. Due to the relatively short 90-minute cycle, they will go through about 5500 cycles per year but as they use a low depth of discharge, usually from 10% to 40% State of Charge (SOC) swing, they are only "partial" cycles which enables lithium-ion to operate successfully in these applications (Borthomieu, 2014).

The second satellite application is the geostationary earth orbit satellites. These satellites stay in one spot relative to a fixed position on the earth and so have a 24-hour cycle, therefore both the discharge rate and the charge rates are very low except for the period when the satellite is eclipsed from the sun which is about a 72 min period of time. With these long charge/discharge cycles, the battery only needs between 1500 and 2000 cycles over the 15- to 20-year life of the satellite (Borthomieu, 2014).

The third and final satellite application for batteries is the high earth orbit and medium earth orbit satellites. Batteries in these applications will require about the same performance requirements as those in the GEO category but will require lifetimes up to about 2500 cycles for up to 15 years of operation or so (Borthomieu, 2014).

There are a couple of things that make satellite batteries unique. First is the need to meet very tough shock and vibration requirements. As these applications will have to be launched via a rocket into space, they must be able to survive this very challenging shock and vibration profile. In addition to this, there is a need to ensure that the batteries are resistant to solar radiation which can negatively impact electronics and material performance. Finally, these batteries must be robust and "bullet proof." By that I mean that it is not possible to fix, repair, or replace them once they have been put into operation. At 23,000 miles above the earth, there is not a local battery repair technician who will be available to make a repair!

Finally, in respect to batteries for satellites, there are two sets of standards that the batteries must be designed and certified to: the European Space Agency and the NASA standards and certifications. For the lithium-ion cells, it is important to test them to the ISO WD17546 standard. The rest of the characterization and testing requirements are very similar to all other lithium-ion batteries and will include electrical performance and characterization testing, abuse testing, and calendar and cycle life testing.

Commercial aviation is also moving toward greater integration of lithium-ion batteries. One of the most recent and highly publicized of these was the introduction of the Boeing 787 Dreamliner aircraft with dual lithium-ion batteries used to power the main and auxiliary power units.

Boeing is perhaps the first commercial company to integrate lithium-ion into their newest commercial planes, but not without incident. After battery fires in a plane in Boston and in Japan, the entire fleet of Dreamliner's was grounded for nearly 3 months while the battery failures were evaluated. During this time, Boeing engineers devised a series of updates to the battery system to prevent failures from spreading outside of the battery itself (Boeing, 2013) (Figure 27).

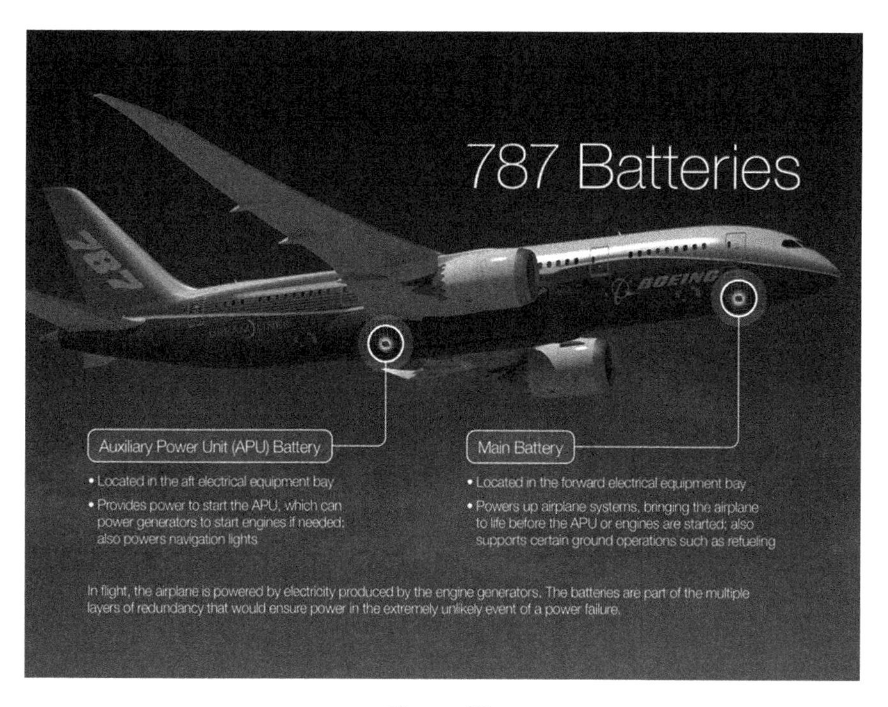

Figure 27
Boeing 787 lithium-ion batteries.

But Boeing is not the only aircraft to use lithium-ion batteries. In fact, many military aviation systems are designed with redundant backup battery systems. Often the main control may be hydraulic, but will have a redundant electrical backup system. This is a very valuable design aspect of lithium-ion batteries for all types of aviation—redundancy. In designing batteries for these types of applications, it is imperative to include double and sometimes even triple levels of redundancy. This will come in several areas, the first is in the combination of series and parallel cells. The second is in the electronics, which will have multiple processors and completely redundant circuits. But the third is the inclusion of the battery itself as a backup to a mechanical control system.

CHAPTER 16

The Future of Lithium-Ion Batteries and Electrification

Someone recently asked me what I thought the battery of 2030 would look like, and how it would operate. This becomes a very interesting question the more that you think about it. The question is not as simple as asking how the technology will evolve or what chemistry will be the most popular; there are also factors such as population growth, the emergence of "mega-cities," and shifting generational trends that we must consider if we attempt to become a "battery futurist." The only thing that I can be absolutely certain of is that the battery of tomorrow will not be the same as the batteries we are using today.

Major Trends

The first question that should be considered is what the society of the future will look like and how that will affect what the regulatory and regional differences that will emerge and drive the technological evolution? In the early 2000s we saw the governments in Europe, Japan, the United States, China, and many others introduce major changes to their fuel economy and CO_2 regulations that began a trend toward more fuel efficient vehicles. This was actually a trend that began as early as the 1970s but had stagnated, at least in the United States. But with continuing global population growth from around 3 billion in 1960 to over 7 billion in 2014 and with forecasts exceeding 8 to 10 billion by 2040–2050 (United Nations, 2014) the demand for new and clean energy solutions will continue to be in high demand. The other factor that we must consider is where the population growth will be centered and from all current forecasts much of this growth will be in large cities. So with world populations continuing to grow, albeit at a somewhat slower rate than it has over the past few decades, and with a consolidation of that growth into cities and the emergence of "mega-cities" with populations of over 20 million, lithium-ion battery technology will have to continue to evolve in order to support this growth trend.

Both the amount and the centralization of this population growth will drive major changes into many technologies in order to manage both the drain on natural resources and impact on the emissions, greenhouse gasses, and pollution. Public functions such as power utilities, for example, have historically been centralized but will evolve into more distributed generation and storage systems. Today power is generated in large coal burning, nuclear, and hydroelectric plants and distributed throughout the grid to their

The Handbook of Lithium-Ion Battery Pack Design. http://dx.doi.org/10.1016/B978-0-12-801456-1.00016-6

customers. But in the future, it may be more likely that power will be both generated and stored locally. We may find rooftops covered in solar panels, wind turbines located in public areas with both using battery storage located in households and in distributed locations throughout these mega-cities.

Transportation will not be missing from this technological transition period. One very interesting and growing trend with the current generation of young adults is that fewer youth aspire to automobile ownership. Instead many are beginning to use car sharing services and living in areas with public transportation. Instead of paying for a vehicle that sits for most of the day and night, they pay for the vehicle only when they need it. In places like Japan we are also seeing a trend toward "personal transportation" which will create a major shift in how consumers expect to use their vehicles. The rest of the world may want to pay attention to the trends in Japan as it is somewhat ahead of the rest of the world in its population trends. As the rate of population growth in Japan has slowed and actually began to move in reverse, they are looking at an aging population in highly populated cities and are looking for solutions that meet the needs of this group. And of course we cannot ignore the growing markets in China and India, where automobile ownership is seen as a status symbol and will continue to experience major growth. But with this increased automotive demand the mega-cities will struggle with growing levels of pollution and lack of available parking. A final thought on the evolution occurring in the transportation segment will be seen in with the upcoming introduction of autonomous vehicles. As more of these vehicles are introduced they will continue to drive changes in the technology that powers them.

Technological Trends

Unfortunately, Moore's Law does not come into effect for batteries! Moore's Law was introduced by Gordon Moore, one of the founders of Intel, back in 1965 when he stated that the amount of transistors on an integrated circuit will double every two years (Intel, 2014). This has proven to be very accurate for integrated circuits, but unfortunately cannot be applied to batteries. If we go back to the introduction of the mass market lithium-ion cell in 1991 battery capacity has only improved about 5–6% per year (Lemkin, 2013; The Week Staff, 2013). But battery technology has not come close to doubling every year and in fact has only improved about eightfold since the first commercial batteries were introduced in 1854.

Coming back to the original question that this chapter started out with, "What does the battery of 2030 look like?" So how does the current battery technology evolve in order to meet these growing needs? Is there really a "game changing" or "disruptive" technology out there?

Unfortunately, we have begun to use the term "disruptive" technology far too frequently and in reference to improvements that are really only minimally disruptive and in fact are really only evolutionary in nature. Let us think about the term disruptive technology for a minute and what it really means. When is technologically disruptive? A technological disruption occurs when some technology entirely changes the way things are now or have been done. The introduction of the automobile was totally disruptive to the horse and wagon and eventually entirely displaced it. The introduction of the personal computer, in a matter of only a few years, almost totally replaced the type writer. The telephone replaced the telegraph, and the cellular phone has virtually replaced the telephone in a matter of years. So with this definition a technology as being disruptive when it replaces and improves on its predecessor, if we look at it in that light is lithium-ion a disruptive technology? Not really, it does not replace the lead acid battery and it does not replace the engine—at least as of this writing! But is it a complementary technology? Absolutely, as it provides significant improvements to the current portfolio of power solutions.

There are a couple of things that could drive lithium-ion into becoming a true disruption: energy density and power density. Lithium-ion battery technology today offers only about 1/10th of the energy of gasoline or diesel fuel. That means that with current chemistries you will not achieve a battery-based solution that can offer the same range as a gas engine vehicle in the same space. But could that change? The story with power density is the same as with energy density. In order to displace a gas engine the battery technology must get smaller (volumetrically) while at the same time increasing in energy and power to a point where it is on par with a liquid-fueled vehicle.

But there is one other aspect that we must consider—cost. Even if someone devises a technology that could offer the same energy content as a liquid fuel vehicle, if the cost does not come down to be in line with the increases it will not become a feasible solution and will not achieve mass market adoption. We are likely to see cost improvements due in large part to some level of standardization and volume growth within the industry. We briefly discussed standardization back in Chapter 7 but as we look into the future and what it may entail we find that some level of standardization will certainly offer benefits for the cost structure. But it should also be flexible enough to allow for continued levels of technological improvement and as technology grows standards must evolve with it.

From an automotive standpoint, once a technology is engineered into a vehicle it is likely to last 5 to 10 years as that is about the life of a standard vehicle architecture. What that means is that the electrified vehicles that are being introduced to the market today were actually designed using batteries that were state of the art three to five years ago. So as I sit here writing this in late 2014, the batteries that are being designed into vehicles today will not become available to the market until 2019 or 2020.

The portable power industry will also help to drive innovation and even to become early adopters of some of the battery technology as we see personal electronics getting smaller and thinner and even wearable. Battery technology will have to evolve along with it.

Future Trends in Battery Technology

So what *could* happen to the battery that could make lithium-ion truly disruptive? Battery chemistry will undoubtedly continue to evolve, but will it ever become truly disruptive? That is difficult to say, lithium is already one of the lightest materials in the periodic table so there is not a lot of opportunity to continue to something outside of a lithium-based technology. So perhaps the disruption will not come by the lithium, but from the other materials.

There are essentially three different parts of the traditional lithium-ion battery that are continuing to be improved: the anode, the cathode, and the electrolytes. By working with these three components, researchers are working to drive up the usable operating voltage of lithium-ion cells, drive up the energy density, and improve the safety and life of the cells.

First we will look at the anodes: silicon and tin anode materials have been receiving significant attention over the past few years. This is because of the potential that they offer for increasing energy density. While the theoretical energy density improvement that can be gained using silicon or tin anodes is in the range of 300% or more, the practical energy density increase is often about one-third of the theoretical number. This will begin to bring lithium-ion into the range of competitiveness with liquid fuels. However, both of these materials also suffer from short life cycle due to the large amount of expansion that they experience during regular charge and discharge cycling which is also in the range of 300%. But that does not mean that we should count it out. There is a lot of work occurring such as looking at nano-materials, coating the silicon or tin with graphite, graphene or other materials, as well as new methods for manufacturing that may offer major improvements in the ability of these chemistries to become mass market capable. We are already beginning to see some of these chemistries emerging in the portable power industry, as we see personal electronics getting smaller and thinner and perhaps even wearable. These applications tend to have much shorter life requirements than transportation or stationary energy storage applications and so become a good ground for the development and testing of these solutions. So as this anode technology continues to evolve, it will eventually replace some if not much of the current graphite-based anode materials.

Cathode materials also continue to see some evolutionary improvements. Some of these come from combining or blending different chemistries in order to get the best parts of both. But there is less opportunity here for making great strides based on current chemistries.

Electrolytes and separators will also continue to see evolutionary changes. Electrolytes are the key to increasing the operating voltages of the lithium-ion cells and continue to be experimented with by adding different additives into the base electrolytes. There is also much work investigating the use of organic electrolytes. Organic electrolytes may end up being safer than the current technologies. Separators are also getting some attention, but this is mostly focused on improving the safety of the cells. Many cell manufacturers are now moving toward the use of a ceramic coated separator as it proves to be less sensitive to penetration testing and can continue to operate effectively at much higher temperatures than traditional polypropylene/ polyethylene separators (Nishi, 2014).

So what other types of improvements can we look at which may hold some truly transformational opportunities for lithium-ion batteries? Perhaps one of the most interesting new technologies that is in development is the solid state battery. A solid state battery is one that simply does not use a liquid electrolyte. Instead the cathode material is built up on a substrate using a deposition method. The separator material is then built upon that again using a deposition methodology. The real benefit of solid state batteries is the improvement in energy density that is achieved by directly depositing one material onto another compared to the traditional lithium-ion which ends up with spaces between the cathode, separator, and anode. And with no liquid electrolyte the safety of the solid state battery is greatly improved. To date most of the solid state lithium-ion batteries have been very small, in the mAh size range. But if developers are able to successfully scale up these cells they may become much more competitive for use in larger applications (Mizuno, Yada, & Iba, 2014).

Another battery technology that is also receiving a lot of research attention today is the lithium-air battery. The lithium-air battery uses a lithium-based anode with a solid electrolyte layer that is "electromechanically coupled" to the oxygen in the air. This offers extremely high energy density, very flat discharge curves, essentially unlimited shelf life as long as it is not exposed to air, low cost, and no environmental issues. However, there are some challenges with it also, the biggest of which is that it is dependent on the environment for the oxygen and it has limited power output capacity (Atwater & Dobley, 2011).

Another technology that has been receiving significant development over the past twenty years or so and should be watched is the fuel cell. Micro fuel cells, automotive fuel cells, and very large fuel cells have all been deployed into the market. These technologies may come to displace some of the lithium-ion battery market if several things come to pass, including the development of a fueling infrastructure and continued cost reduction. The fuel cell is essentially an electricity generator that may begin to grow in many of the same markets that lithium-ion batteries are being introduced into.

Another set of technologies that is growing are the supercapacitors and ultracapacitors. These technologies have historically only been adequate for extremely short duration bursts of very high power. However, current research is working to increase the energy density of these technologies, thereby making them much more competitive with traditional batteries. Additionally, as their manufacturing process is very similar to lithium-ion battery manufacturing, we may begin to see cell manufacturers adding these technologies to their product portfolios.

Conclusion

There are many potential near-term improvements that are already in the works for lithium-ion batteries and which will offer incremental improvements to the current cells. The cells that power the world in 2030 will probably not even begin to get validated for high volume applications until about 2025 or so, maybe a year or two earlier. Costs will continue to come down as volumes continue to increase and some of these incremental technological improvements that I mentioned will come to pass which will also help to drive down costs. And if we do it right the next generation of technologies may come in the same form factors that we are building today in order to ease the transition from one technology to the next. But I believe that it will be difficult to achieve some of the cost targets that have been set out by groups such as the U.S. DOE and the USABC of $100–$150/kWh. With 18650 cells today being built at a rate of nearly 700 million units per year and have bottomed out at prices in the range of $170–$220/kWh what would it take for these new large format cells to reach those same levels? Due to the size of the cells they are not likely to reach the same levels of volume, so that means that it must be a technological shift that will drive down the costs. Which likely means it will be one of the ones that we discussed here.

Growth in the mega-cities will drive a change toward cleaner energy generation and storage technologies. The battery-based energy storage system may very likely become an essential part of being able to provide uninterrupted power to these cities. And transportation will be forced to electrify in order to reduce the impacts of emissions on these large population centers.

And to be perfectly honest, the battery of 2030 and beyond is probably still in a test tube today if it is even a concept. It is more than likely that the battery even today is yet to be developed!

References

Abuelsamid, S. (August 18, 2009). *Toxco gets $9.5 million DOE grant for battery recycling*. Retrieved December 14, 2014, from Autobloggreen.com: http://green.autoblog.com/2009/08/18/toxco-gets-9-5-million-doe-grant-for-battery-recycling/.

Advanced Lead Acid Battery Consortium. (2011). *Advanced Lead Acid Battery Consortium*. Retrieved from http://www.alabc.org/.

Akhil, A. A., Huff, G., Currier, A. B., Kaun, B. C., Rastler, D. M., Chen, S. B., et al. (2013). *DOE/EPRI 2013 Electricity storage handbook in collaboration with NRECA*. Albuquerque, New Mexico: Sandia National Laboratories.

ALCAD. (2010). *Battery sizing*. Retrieved August 6, 2014, from Alcad.com: http://alcad.com/tools/battery-sizing.

American Society for Quality. (2014). *Failure mode effects analysis (FMEA)*. Retrieved from ASQ.org: http://asq.org/learn-about-quality/process-analysis-tools/overview/fmea.html.

Andrea, D. (2010). *Battery management systems for large lithium-ion battery packs*. Boston, MA: Artech House.

ANSYS. (2014). Retrieved August 8, 2014, from ANSYS: http://ansys.com/.

Arendt, S. (October 26, 2006). *Sony battery recall costs $429 million*. P. Mag, Producer. Retrieved 2014, from http://www.pcmag.com/article2/0,2817,2040936,00.asp.

Ashby, D. (2009). *Electrical engineering 101* (2nd ed.). Amsterdam: Newnes.

Association Française de Normalisation. (2009). *European Standard NF EN 15194: Electrically power assisted cycles*. Association Française de Normalisation (AFNOR — French standard institute).

ATC New Technologies. (2014). *High voltage battery packs - repair, remanufacture and repurpose*. Retrieved December 14, 2014, from ATC Drivetrain: http://www.atcdrivetrain.com/services/repair-batteries.php.

Atwater, T. B., & Dobley, A. (2011). Metal/Air batteries. In T. B. Reddy (Ed.), *Linden's handbook of batteries* (4th ed.) (pp. 33.1–33.58). New York: McGraw-Hill.

Aurbach, D., Talyosef, Y., Markovsky, B., Markevich, E., Zinigrad, E., Asraf, L., et al. (2004). Design of electrolyte solutions for Li and Li-ion batteries: a review. *Electrochimica Acta*, *50*, 247–254.

Automotive Industry Action Group. (2014). *AIAG training services: Failure mode and effects analysis (FMEA) overview*. Retrieved from aiag.org: https://www.aiag.org/staticcontent/education/trainingindex.cfm?classcode=FMEA.

Bailo, C. (July 2012). Retrieved August 3, 2014, from MyNissanLeaf.com: http://www.mynissanleaf.com/assets/An%20open%20letter%20to%20Nissan%20LEAF%20owners%20from%20Carla%20Bailo_FINAL.pdf.

Barrett, J. P. (1894). *Electricity at the columbian exposition*. Chicago: R.R. Donnelley & Sons.

Battery Council International. (2013). *Battery Council International (BCI)*. Retrieved from http://batterycouncil.org/.

Behr. (July 20, 2010). *Behr's cooling technology boosts electric vehicles and gasoline engines*. Retrieved December 2, 2014, from businesswire.com: http://www.businesswire.com/news/home/20100720006986/en/Behr%E2%80%99s-Cooling-Technology-Boosts-Electric-Vehicles-Gasoline#.VH5wB9EtDIU.

Boeing. (2013). *Batteries and advanced airplanes: 787 electrical system*. Retrieved December 19, 2014, from Boeing.com: http://787updates.newairplane.com/787-Electrical-Systems/Batteries-and-Advanced-Airplanes#.

Borthomieu, Y. (2014). Satellite lithium-ion batteries. In G. Pistoia (Ed.), *Lithium-ion batteries: Advances and applications* (pp. 311–314). Amsterdam: Elsevier.

Brill, J. N. (2011). Nickel-hydrogen batteries. In T. B. Reddy (Ed.), *Linden's handbook of batteries* (pp. 24.1–24.28). New York: McGraw-Hill.

Buchmann, I. (2011). *Batteries in a portable world: A handbook on rechargeable batteries for non-engineers* (3rd ed.). Richmond, BC: Cadex Electronics Inc.

California Air Resource Board. (2013). *Facts about California's commercial vehicle idling regulations.* Sacramento: ARB. Retrieved September 1, 2014, from www.arb.ca.gov/msprog/truck-idling/factsheet.pdf.

California Air Resource Board. (2014). *Zero emissions vehicle (ZEV) program.* Retrieved from California Environmental Protection Agency Air Resources Board: http://www.arb.ca.gov/msprog/zevprog/zevprog.htm.

Carter, S. (July 26, 2012). Nissan pens open letter to LEAF customers affected by battery capacity loss. *Nissan LEAF Electric Car News.* Retrieved August 3, 2014, from http://nissan-leaf.net/2012/07/26/nissan-pens-open-letter-to-leaf-customers-affected-by-battery-capacity-loss/.

China Automotive Technology and Research Center. (n.d.). *About CATARC.* Retrieved on August 31, 2014, from China Automotive Technology & Research Center: http://tv.catarc.ac.cn/ac_en/about/intro/webinfo/2007/12/1196727076728291.htm

Ciferri, L. (September 25, 2014). *Renault-Nissan rethinks EV battery strategy.* Retrieved September 27, 2014, from Automotive News Europe: http://europe.autonews.com/article/20140925/ANE/140929945/renault-nissan-rethinks-ev-battery-strategy.

Clark, N. N., Zhen, F., Wayne, W. S., & Lyons, D. W. (2007). *Transit bus life cycle cost and year 2007 emissions estimation. West Virginia University, Center for Alternative Fuels, Engines & Emissions Dept. of Mechanical & Aerospace Engineering.* Morgantown, WV: Federal Transit Administration, U.S. Department of Transportation. Retrieved August 31, 2014, from www.fta.dot.gov/documents/WVU_FTA_LCC_Final_Report_07-23-2007.pdf.

Cobasys. (January 10, 2006). *Cobasys to provide battery systems for saturn VUE green line hybrid SUV.* Retrieved from EV World.com: http://evworld.com/news.cfm?newsid=10652.

Cobb, J. (August 22, 2014). *GM hydrogen fuel cell vehicle update.* Retrieved from GM-Volt http://gm-volt.com/2012/08/22/gm-hydrogen-fuel-cell-vehicle-update/.

Cobb, J. (May 10, 2013). *A major shift toward micro hybrids on the horizon.* Retrieved August 10, 2014, from hybridcars.com: http://www.hybridcars.com/a-major-shift-toward-micro-hybrids-on-the-horizon/.

COMSOL. (2014). Retrieved August 8, 2014, from COMSOL: http://www.comsol.com/.

Dahn, J., & Ehrlich, G. M. (2011). Lithium-ion batteries. In T. B. Reddy (Ed.), *Linden's handbook of batteries* (pp. 26.1–26.79). New York: Mcgraw-Hill.

DesignBoom. (March 11, 2013). *Toyota i Road electric personal mobility vehicle.* Retrieved August 10, 2014, from DesignBoom: http://www.designboom.com/technology/toyota-i-road-electric-personal-mobility-vehicle/.

Dhameja, S. (2002). *Electric vehicle battery systems.* Boston: Newnes.

Dlegs Hybrid Cars. (February 28, 2011). *2012 Buick regal to offer gas savings in the form of eAssist mild hybrid technology.* Retrieved from dlegs.com: http://www.dlegs.com/tag/regal-e-assist-lithium-ion-battery-cost/.

DNV GL. (2013). *DNV GL guideline for large maritime battery systems.* DNV GL.

DNV-GL. (2014). *Det Norske Veritas.* Retrieved from http://dnvgl.com/.

Electric Drive Transportation Association. (n.d.). Electric drive transportation association. Retrieved on August 31, 2014, from http://electricdrive.org/

Energy Storage Association. (2014). *Community energy storage.* Retrieved November 9, 2014, from Energy Storage Association: http://energystorage.org/energy-storage/technology-applications/community-energy-storage.

Erbacher, J. K. (2011). Industrial and aerospace Nickel-Cadmium batteries. In T. B. Reddy (Ed.), *Linden's handbook of batteries* (pp. 19.1–19.23).

EUROBAT. (2010). *EUROBAT Home page.* Retrieved from EUROBAT: http://www.eurobat.org/.

European Council for Automotive R&D. (2014). *European council for automotive R&D.* Retrieved from European Council for Automotive R&D: http://www.eucar.be/.

Felton, N. (2008). Consumption spreads faster today. *The New York Times.* Retrieved November 2, 2014, from http://www.nytimes.com/imagepages/2008/02/10/opinion/10op.graphic.ready.html.

FreedomCAR Program Electrochemical Energy Storage Team. (2003). *FreedomCAR battery test manual for power-assist hybrid electric vehicles.* United States Deparment of Energy. Idaho National Engineering & Environmental Laboratory. Retrieved 2014.

General Motors. (March 3, 2010). *GM Unveils EN-V concept: A vision for future urban mobility*. Retrieved August 10, 2014, from GM Auto Shows: News: http://media.gm.com/autoshows/Shanghai/2010/public/cn/en/env/news.detail.html/content/Pages/news/cn/en/2010/March/env01.html.

General Motors. (November 14, 2012). *GM, ABB demonstrate Chevrolet Volt battery reuse unit*. Retrieved August 16, 2014, from Genetral Motors: New: http://media.gm.com/media/us/en/gm/news.detail.html/content/Pages/news/us/en/2012/Nov/electrification/1114_reuse.html.

Gordon-Bloomfield, N. (August 26, 2013). Nissan testing new battery pack to address problems with LEAF in hot weather. *Plugincars*. Retrieved August 3, 2014, from http://www.plugincars.com/nissan-testing-new-battery-pack-leaf-128088.html.

Gordon-Bloomfield, N. (September 28, 2012). Nissan buys back leaf electric cars under Arizona Lemon Law. *Green Car Reports*. Retrieved August 3, 2014, from http://www.greencarreports.com/news/1079475_nissan-buys-back-leaf-electric-cars-under-arizona-lemon-law.

Green Car Congress. (December 9, 2009). *2010 Prius plug-in hybrid makes North American debut at Los Angeles auto show; first li-ioon battery traction battery developed by Toyota and PEVE*. Retrieved August 17, 2014, from Green Car Congress: http://www.greencarcongress.com/2009/12/prius-phv-20091202.html.

Green Car Congress. (October 24, 2013). *Governors of 8 states sign MoU to put 3.3M zero-emission vehicles on roads by 2025; 15% of new vehicle sales*. Retrieved November 9, 2014, from Green Car Congress: http://www.greencarcongress.com/2013/10/governors-of-8-states-sign-mou-to-put-33m-zero-emission-vehicles-on-roads-by-2025-15-of-new-vehicle-sales.html.

Greenfield, R. (March 20, 2012). *Meet the little orange robots making amazon's warehouses more humane*. Retrieved September 28, 2014, from The Wire: http://www.thewire.com/technology/2012/03/meet-little-orange-robots-making-amazons-warehouses-more-humane/50094/.

Gross, L. P., & Snyder, T. R. (2005). *Philadelphia's 1876 centennial exhibition*. Arcadia Publishing.

Halderman, J. D., & Martin, T. (2009). *Hybrid and alternative fuel vehicles*. Upper Saddle River: Prentice Hall.

Herzberg, F., Mausner, B., & Snyderman, B. B. (1959). *The motivation to work*. New York: John Wiley & Sons.

Hoehner Research & Consulting Group GmbH. (2013). *IBESA Home*. Retrieved from International Battery and Energy Storage Alliance: http://www.ibesalliance.org/energy-storage-home.html.

Holmes, J. (October 14, 2014). *Toyota global hybrid sales crest 7 million units*. Retrieved November 23, 2014, from Automobile: http://www.automobilemag.com/features/news/1410-toyota-global-hybrid-sales-crest-7-million-units/.

Honda. (February 24, 2010). *Honda 3R-C concept world debut at Geneva international motor show 2010*. Retrieved August 10, 2014, from Honda Worldwide Home: World News: http://world.honda.com/news/2010/4100224Geneva-Motor-Show/.

HQEW.net. (August 19, 2012). *Working of a capacitor. HQEW.net*. Retrieved from http://circuit-diagram.hqew.net/Working-of-a-Capacitor_9752.html.

IEEE. (2014). *1115-2000-IEEE recommended practice for sizing Nickel-Cadmium batteries for stationary applications*. Retrieved from IEEE Standards Association: http://standards.ieee.org/findstds/standard/1115-2000.html.

IEEE. (2014). *485-2010-IEEE recommended practice for sizing lead-acid batteries for stationary applications*. Retrieved from IEEE Standards Association: http://standards.ieee.org/findstds/standard/485-2010.html.

IEEE. (2014). *IEEE: Advancing technology for humanity*. Retrieved from http://www.ieee.org/index.html?WT.mc_id=hpf_logo.

Intel. (2014). *Moore's law inspires intel innovation*. Retrieved December 19, 2014, from Intel.com: http://www.intel.com/content/www/us/en/silicon-innovations/moores-law-technology.html.

International Maritime Organization. (2014). *Hisotry of MARPOL*. Retrieved December 19, 2014, from IMO: International Maritime Organization: http://www.imo.org/KnowledgeCentre/ReferencesAndArchives/HistoryofMARPOL/Pages/default.aspx.

International Organization for Standardization. (2014). *About us*. Retrieved from ISO: http://www.iso.org/iso/home/about.htm.

Japan Echo Inc. (July 23, 2003). *Fuel cells for the home*. Retrieved September 28, 2014, from Trends in technology: Science & technology: http://web-japan.org/trends/science/sci030723.html.

Jonnes, J. (2004). *Empires of light: Edison, Tesla, Westinghouse, and the race to electrify the world*. New York: Random House Trade Paperbacks.

King, D. (January 12, 2013). Leaf owners not happy with Nissan's response to battery problems. *Autobloggreen*. Retrieved August 3, 2014, from http://green.autoblog.com/2013/01/12/leaf-owners-not-happy-with-nissans-response-to-battery-problems/.

Kirsch, D. A. (2000). *The electric vehicle and the Burden of history*. Newark: Rutgers University Press.

Klayman, B. (October 3, 2013). *Tesla grapples with impact of battery fire in U.S.* Retrieved October 4, 2014, from Reuters.com: http://www.reuters.com/article/2013/10/03/us-autos-tesla-fire-idUSBRE9920SX20131003.

Larsen, E. (2003). *The devil in the white city: Murder, magic and madness at the fair that changed America*. New York, NY: Crown.

LeGault, M. (April 1, 2013). *Chevrolet Volt battery pack: Rugged but precise*. Retrieved December 7, 2014, from Composites Technology: http://www.compositesworld.com/articles/chevy-volt-battery-pack-rugged-but-precise.

Lemkin, J. M. (June 1, 2013). *Why haven't there been improvements to battery technology recently?* Retrieved December 19, 2014, from Quora.com: https://www.quora.com/Why-havent-there-been-improvements-to-battery-technology-recently.

Lowy, J. (November 26, 2011). *Chevy volt battery catches fire, government investigates general motors' electric car*. Retrieved October 4, 2014, from Huffington Post.com: http://www.huffingtonpost.com/2011/11/26/chevy-volt-battery-fire-electric-car-general-motors_n_1114193.html.

MathWorks. (2014a). *Matlab*. Retrieved August 8, 2014, from MathWorks: http://www.mathworks.com/products/matlab/.

MathWorks. (2014b). *Simulink*. Retrieved August 8, 2014, from MathWorks: http://www.mathworks.com/products/simulink/.

Maxim Integrated. (October 31, 2007). *APPLICATION NOTE 4126: Understanding the IP (ingress protection) ratings of iButton data loggers and capsule*. Retrieved December 7, 2014, from MaximIntegrated.com: http://www.maximintegrated.com/en/app-notes/index.mvp/id/4126.

Mizuno, F., Yada, C., & Iba, H. (2014). Solid state lithium-ion batteries for electric vehicles. In G. Pistoia (Ed.), *Lithium-ion batteries: Advances and applications* (pp. 273–291). Amsterdam: Elsevier.

Mook, N. (2006). *Sony battery recall cost: $250 million*. Retrieved November 27, 2014, from betanews.com: http://betanews.com/2006/08/25/sony-battery-recall-cost-250-million/.

NAATBAtt. (2014). *NAATBatt: National alliance for advanced technology batteries*. Retrieved 2014, from http://naatbatt.org/.

Nanosteel. (2014). *AHSS sheet: New class of steel for lightweighting automobiles*. Retrieved December 7, 2014, from The Nanosteel Company: http://nanosteelco.com/products/sheet-steel/ahss-sheet.

National Electrical Manufacturers Association. (November 2005). *NEMA enclosure types*. Retrieved December 7, 2014, from National Electrical Manufacturers Association: http://www.nema.org/prod/be/enclosures/upload/NEMA_Enclosure_Types.pdf.

National Highway Traffic Safety Administration. (2014). *CAFE - Fuel economy*. Retrieved November 9, 2014, from NHTSA National Highway Traffic Safety Administration: http://www.nhtsa.gov/fuel-economy.

National Renewables Energy Laboratory. (August 14, 2013). *Modeling & simulation - computer-aided engineering for electric-drive vehicle batteries*. Retrieved August 6, 2014, from NREL Vehicles & Fuels Research: http://www.nrel.gov/vehiclesandfuels/energystorage/caebat.html.

Nishi, Y. (2014). Past, present and future of lithium-ion batteries: can new technologies open up new horizons? In G. Pistoia (Ed.), *Lithium-ion batteries: Advances and applications* (pp. 21–39). Amsterdam: Elsevier.

Nissan. (2011). *2012 LEAF owner's manual. The EV (Electric Vehicle) System, EV-2*. Retrieved August 3, 2014, from https://owners.nissanusa.com/nowners/navigation/manualsGuide.

Nissan. (February 9, 2014). *Nissan leaf batteries to get 'second life' as grid storage*. Retrieved August 16, 2014, from ev world.com/Press Release: http://evworld.com/news.cfm?rssid=32325.

O'Dell, J. (June 30, 2014). *Car news*. Retrieved August 3, 2014, from Edmunds.com: http://www.edmunds.com/car-news/nissan-prices-leaf-battery-replacement-at-5500.html.

Pacific Northwest National Laboratory. (2014). *Inventory of safety-related codes and standards for energy storage systems: With some experiences related to approval and acceptance*. Richland, Washington: Pacific Northwest National Laboratory. Retrieved 2014.

Portable Rechargeable Battery Association. (2014). *About PRBA*. Retrieved from PRBA The Rechargeable Battery Association: http://www.prba.org/about-prba/.

Rahn, C. D., & Wang, C.-Y. (2013). *Battery systems engineering*. West Sussex, UK: John Wiley & Sons.

Reddy, T. B. (2011). *Linden's handbook of batteries* (4th ed.). New York: McGraw Hill.

ReliaSoft. (2014). Design for reliability: overview of the process and applicable techniques. *Reliasoft's Reliability Edge*, *8*(2), 1–6. Retrieved February 2014, from Reliability Edge: http://www.reliasoft.com/newsletter/v8i2/reliability.htm.

Rhodes, M. (November 21, 2014). *Urban planning ideas for 2030, when billions will live in megacities*. Retrieved December 18, 2014, from Wired.com: http://www.wired.com/2014/11/urban-planning-ideas-2030-billions-will-live-megacities/.

Rousch, M. (March 26, 2012). *A123 recalling $55M in EV batteries made in Livonia*. Retrieved from CBS Detroit: http://detroit.cbslocal.com/2012/03/26/a123-recalling-55m-in-ev-batteries-made-in-livonia/.

Runte, G. (April 11, 2014). *Fuel cells in the (Japanese) home!*. Retrieved August 28, 2014, from Worthington Sawtelle: http://worthingtonsawtelle.com/fuel-cells-japanese-home/.

SAE International. (2014). *Vehicle electrification standards*. Retrieved from SAE International: http://www.sae.org/standardsdev/vehicleelectrification.htm.

Sahraie, E., Meier, J., & Wierzbicki, T. (2014). Characterizing and modeling mechanical properties and onset of short circuit for three types of lithium-ion pouch cells. *Journal of Power Sources*, *247*, 503–516.

Sandia National Laboratories. (2006). *FreecomCAR electrical energy storage system abuse testing manual for electricand hybrid electric vehicle applications*. Albuquerque: Sandia National Laboratories. Retrieved February 2014.

Santayana, G. (1905). *The life of reason: (Vol. 1) Or, the phases of human progress. "Reason in common sense"*. (p. 284). Prometheus Books.

Santos, A. (January 16, 2013). *FAA grounds all US Boeing 787 dreamliners after second lithium ion battery failure*. Retrieved October 4, 2014, from engadget: http://www.engadget.com/2013/01/16/faa-grounds-all-us-boeing-787-dreamliners-lithium-ion-battery/.

Schiffer, M. B. (1994). *The electric automobile in America: Taking charge*. Washington and London: Smithsonian Books.

Seifer, M. (1996). *Wizard: The Life and Times of Nikola Tesla: Biography of a Genius*. New York, NY: Citadel Press.

Society for Automotive Engineers. (2014). *Society for automotive engineers*. Retrieved from http://www.sae.org/.

Society for Automotive Engineers. (2014). *Society for automotive engineers*. Retrieved from SAE International: http://www.sae.org/.

Sony. (August 25, 2006). *Statement regarding Sony's support of Apple's recall of lithium-ion battery packs used in Apple notebook computers*. Retrieved August 31, 2014, from Sony: Corporate Info: http://www.sony.net/SonyInfo/News/Press/200608/06-0825E/index.html.

Spiers, D. (2014). CEO, Spiers New Technologies. (J. Warner, Personal Communication)

Standardization Administration of the People's Republic of China. (2014). *Standardization Administration of the People's Republic of China*. Retrieved from Standardization Administration of the People's Republic of China: http://www.sac.gov.cn/sac_en/.

Sweet, B. (October 25, 2013). *California's first-in-nation energy storage mandage*. Retrieved August 31, 2014, from IEE Spectrum: http://spectrum.ieee.org/energywise/energy/renewables/californias-firstinnation-energy-storage-mandate.

Sybesma's Electronics. (2014). *Lithium-ion battery repair*. Retrieved December 14, 2014, from Sybesma's Electronics: http://sybesmas.com/site/lithium-ion-battery-repair/.

The Phillips Museum of Art. (2008). *Electrical battery" of Leyden jars, 1760–1769*. Franklin & Marshall College. Retrieved November 3, 2014, from The Benjamin Franklin Tercentenary: http://www.benfranklin300.org/frankliniana/result.php?id=72&sec=0.

The Week Staff. (May 5, 2013). *The search for a better battery*. Retrieved December 19, 2014, from The Week.com: http://theweek.com/article/index/243576/the-search-for-a-better-battery.

Toyota Prius Battery.com. (2013). *Prius battery specifications*. Retrieved August 17, 2014, from Toyota Prius Battery: http://toyotapriusbattery.com.

TrendForce. (May 6, 2013). *Reduced cell cost suggests the upcoming era of large capacity cells*. Retrieved from EnergyTrend: http://www.energytrend.com/price/20130506-5180.html.

UC San Diego. (2014). *2nd life electric vehicle battery project*. Retrieved December 14, 2014, from UC San Diego Resource Management & Planning: http://rmp.ucsd.edu/strategic-energy/storage/2-life.html.

UL Prospector. (2014). *UL 94 flame rating*. Retrieved December 7, 2014, from ULProspector.com: http://www2.ulprospector.com/property_descriptions/UL94.asp.

Umicore. (2014). *Umicore battery recycling*. Retrieved December 14, 2014, from Umicore Battery Recycling: http://www.batteryrecycling.umicore.com/UBR/.

Underwriter's Laboratory. (2014). *Batteries: Services for small batteries*. Retrieved 2014, from Underwriter's Laboratory: http://industries.ul.com/high-tech/batteries.

Underwriter's Laboratory. (2014). *Catalog of standards: UL 2580 batteries for use in electric vehicles*. Retrieved from Underwriter's Laboratory: http://www.ul.com/global/eng/pages/solutions/standards/accessstandards/catalogofstandards/standard/?id=2580_2.

Underwriter's Laboratory. (2014). *Our business*. Retrieved from UL: http://ul.com/?noredirect.

United Nations. (2009). *Recommendations on the transport of dangerous goods: Manual of tests and criteria*. New York and Geneva: United Nations.

United Nations. (2013). *Recommendations on the transport of dangerous goods: manual of tests and criteria* Amendment 1 (5th revised ed.). New York and Geneva: United Nations. Retrieved from. http://www.unece.org/trans/danger/publi/manual/Manual%20Rev5%20Section%2038-3.pdf.

United Nations. (July 10, 2014). *World's population increasingly urban with more than half living in urban areas*. Retrieved December 18, 2014, from United Nations: http://www.un.org/en/development/desa/news/population/world-urbanization-prospects-2014.html.

United States Advanced Battery Consortium. (1999). *United States advanced battery consortium electrochemical storage system abuse test procedure manual*.

USCAR. (2014). *USCAR, United States Council for Automotive Research*. Retrieved 2014, from USABC, United States Advanced Battery Consortium: http://www.uscar.org/guest/view_team.php?teams_id=12.

Verband der Automobilindustrie. (2014). *Home*. Retrieved from VDA: Verband der Automobilindustrie: http://www.vda.de/en/index.html.

Vink, D. (November 19, 2012). *GM cars use plastics in power storage*. Retrieved December 7, 2014, from Plastics News: http://www.plasticsnews.com/article/20121119/NEWS/311199979/gm-cars-use-plastics-in-power-storage.

Voelcker, J. (May 15, 2014). *2015 Chevrolet Spark EV switches battery cells; 82-Mile range remains*. Retrieved September 27, 2014, from www.Greencarreports.com: http://www.greencarreports.com/news/1092094_2015-chevrolet-spark-ev-switches-battery-cells-82-mile-range-remains.

Voelcker, J. (December 12, 2011). *Aptera collapse: How & why it happened, a complete chronology*. Retrieved December 18, 2014, from GreenCarReports.com: http://www.greencarreports.com/news/1070490_aptera-collapse-how-why-it-happened-a-complete-chronology.

Voelcker, J. (March 26, 2012). *A123 systems to recall electric-car battery packs for Fisker, Others*. Retrieved from Green Car Reports: http://www.greencarreports.com/news/1074491_a123-systems-to-recall-electric-car-battery-packs-for-fisker-others.

Voelker, P. (June/July 2014). A closer look at how batteries fail. *Charged: Electric Vehicles Magazine*, *14*, 20–25.

Warner, J. (2014). Lithiuum-ion battery packs for EVs. In G. Pistoia (Ed.), *Handbook of Lithium-ion battery applications* (pp. 127–150). Amsterdam: Elsevier.

Xia, Y., Wierzbicki, T., Sahraei, E., & Zhang, X. (December 1, 2014). Damage of cells and battery packs due to ground impact. *Journal of Power Sources*, *267*, 78–97.

Zhang, S. S. (2006). A review on electrolyte additives for lithium-ion batteries. *Journal of Power Sources*, *162*, 1379–1394.

Zia, Y., Wierzbicki, T., Sahraei, E., & Zhang, X. (December 1, 2014). Damage of cells and battery packs due to ground impact. *Journal of Power Sources*, 78–97.

Appendix A: USABC 12-V Stop/Start Battery Pack Goals

End of Life Characteristics	Units	Target	
		Under Hood	Not Under Hood
Discharge Pulse, 1 s	kW	6	
Max Discharge Current, 0.5 s	A	900	
Cold cranking power at −30 °C (three 4.5-s pulses, 10 s rest between pulses at minimum SOC)	kW	6 KW for 0.5 s followed by 4 kW for 4 s	
Minimum voltage under cold crank	Vdc	8.0	
Available energy (750 W accessory load power)	Wh	360	
Peak recharge rate, 10 s	kW	2.2	
Sustained recharge rate	W	750	
Cycle life, every 10% life RPT with cold crank at minimum SOC	Engine starts/ miles	450,000/150,000	
Calendar life at 30 °C, 45 °C if under hood	Years	15 at 45 °C	15 at 30 °C
Minimum round trip energy efficiency	%	95	
Maximum allowable self-discharge rate	Wh/day	2	
Peak operating voltage, 10 s	Vdc	15.0	
Sustained operating voltage—maximum	Vdc	14.6	
Minimum operating voltage under autostart	Vdc	10.5	
Operating temperature range (available energy to allow 6-kW, 91-s pulse)	°C	−30 °C to +75 °C	−30 °C to +52 °C
30 °C to 52 °C	Wh	360 (to 75 °C)	360
0 °C	Wh	180	
−10 °C	Wh	108	
−20 °C	Wh	54	
−30 °C	Wh	36	
Survival temperature range	°C	−46 °C to +100 °C	−46 °C to +66 °C
Maximum system weight	kg	10	
Maximum system volume	L	7	
Maximum selling price (at 250,000 units/year)	$ USD	$220	$180

SOC, state of charge.

Appendix B: USABC 48-V Battery Pack Goals

Characteristics	Units	Target
Peak pulse discharge power (10 s)	kW	9
Peak pulse discharge power (1 s)	kW	11
Peak regen pulse power (5 s)	kW	11
Cold cranking power at −30 °C (three 4.5-s pulses, 10-s rest between pulses at minimum SOC)	kW	6 kW for 0.5 s followed by 4 kW for 4 s
Accessory load (2.5-min duration)	kW	5
Available energy for cycling	Wh	105
CS 48-V HEV cycle life	Cycles/MWh	75,000/21
Calendar life, 30 °C	Years	15
Minimum round trip energy efficiency	%	95
Maximum self-discharge	Wh/day	1
Maximum operating voltage	Vdc	52
Minimum operating voltage	Vdc	38
Minimum voltage during cold crank	Vdc	26
Unassisted operating temperature range (Power available to allow 5-s charge and 1-s discharge pulse) at minimum and maximum operating SOC and voltage	°C	−30 °C to +52 °C
30–52 °C	kW	11
0 °C	kW	5.5
−10 °C	kW	3.3
−20 °C	kW	1.7
−30 °C	kW	1.1
Survival temperature range	°C	−46 °C to +66 °C
Maximum system weight	kg	≤8
Maximum system volume	L	≤8
Maximum selling price (at 250,000 units/year)	$ USD	$ 275

SOC, State of charge; HEV, hybrid electric vehicle.

Appendix C: USABC HEV Battery Pack Goals

Characteristics	Units	Power Assist Target (Minimum)	Power Assist Target (Maximum)
Peak pulse discharge power (10 s)	kW	25	40
Peak regen pulse power (5 s)	kW	20 (55-Wh pulse)	35 (95-Wh pulse)
Cold cranking power at −30 °C (three 2-s pulses, 10-s rest between pulses at minimum SOC)	kW	5	7
Total available energy (over DOD range where power goals are met)	kWh	0.3 (at C/1 rate)	0.5 (at C/1 rate)
Cycle life for specified SOC increments	Cycles	300,000 25-Wh cycles (7.5 MWh)	300,000 50-Wh cycles (15 MWh)
Calendar life	Years	15	15
Minimum round trip energy efficiency	%	90 (25-Wh cycles)	90 (50-Wh cycles)
Maximum self-discharge	Wh/day	50	50
Maximum operating voltage	Vdc	≤400	
Minimum operating voltage	Vdc	≥ (0.55 × Vmax)	> (0.55 × Vmax)
Unassisted operating temperature range (Power available to allow 5-s charge and 1-s discharge pulse) at minimum and maximum operating SOC and voltage	°C	−30 °C to +52 °C	
Survival temperature range	°C	−46 °C to +66 °C	
Maximum system weight	kg	40	60
Maximum system volume	L	32	45
Maximum selling price (at 250,000 units/year)	$ USD	$ 500	$ 800

SOC, State of charge; DOD, Depth of Discharge.

Appendix D: USABC PHEV Battery Pack Goals

Characteristics	Units	PHEV-20 Target	PHEV-40 Target	xEV-50 Target
Commercialization time frame		2018	2018	2020
All electric range	Miles	20	40	50
Peak pulse discharge power (10 s)	kW	37	38	100
Peak pulse discharge power (2 s)	kW	45	46	110
Peak regen pulse power (10 s)	kW	25	25	60
Cold cranking power at −30 °C (three 2-s pulses, 10-s rest between pulses at minimum SOC)	kW	7	7	7
Available energy for charge depleting (CD) mode	kWh	5.8	11.6	14.5
Available energy for charge sustaining (CS) mode	kWh	0.3	0.3	0.3
Charge depleting cycle life/Discharge throughput	Cycles/MWh	5000/29	5000/58	5000/72.5
Charge sustaining HEV cycle life, 50 Wh profile	Cycles	300,000	300,000	300,000
Calendar life, 30 °C	Years	15	15	15
Minimum round trip energy efficiency	%	90	90	90
Maximum self-discharge	%/month	<1	<1	<1
Maximum operating voltage	Vdc	420	420	420
Minimum operating voltage	Vdc	220	220	220
System recharge rate at 30 °C	kW	3.3 (240 V/16 A)	3.3 (240 V/16 A)	6.6 (240 V/32 A)

—Cont'd

Characteristics	Units	PHEV-20 Target	PHEV-40 Target	xEV-50 Target
Unassisted operating temperature range (Power available to allow 5-s charge and 1-s discharge pulse) at minimum and maximum operating SOC and voltage	°C	−30 °C to +52 °C	−30 °C to +52 °C	−30 °C to +52 °C
30–52 °C	%	100	100	100
0 °C	%	50	50	50
−10 °C	%	30	30	30
−20 °C	%	15	15	15
−30 °C	%	10	10	10
Survival temperature range	°C	−46 °C to +66 °C	−46 °C to +66 °C	−46 °C to +66 °C
Maximum system weight	kg	70	120	150
Maximum system volume	L	47	80	100
Maximum selling price (at 100,000 units/year)	$USD	$2200	$3400	$4250

SOC, State of charge; PHEV, plug-in hybrid electric vehicle

Appendix E: USABC EV Battery Pack Goals

End of Life Characteristics at 30 °C	Units	System Level Target	Cell Level Target
Commercialization time frame		2020	2020
Peak discharge power density (30-s pulse)	W/L	1000	1500
Peak specific discharge power (30-s pulse)	W/kg	470	700
Peak specific regen pulse (10 s)	Wkg	200	300
Peak current, 30 s	A	400	400
Useable energy density at C/3 discharge rate	Wh/L	500	750
Useable specific energy at C/3 discharge rate	Wh/kg	235	350
Cycle life	Cycles	1000	1000
Calendar life, 30 °C	Years	15	15
Maximum self-discharge	%/month	<1	<1
Maximum operating voltage	Vdc	420	N/A
Minimum operating voltage	Vdc	220	N/A
System recharge time	Hours	<7 h, J1772	
Operating environment	°C	−30 °C to +52 °C	
Unassisted operating at low temperature	%	>70% useable energy at C/3 discharge rate at −20 °C	
Survival temperature range, 24 h	°C	−46 °C to +66 °C	
Maximum selling price (at 100,000 units/year)	$ USD/kWh	$ 125	$ 100

Index

Note: Page numbers followed by "f" and "t" indicate figures and tables respectively.

9780128014561